Reviewing Intermediate-Level Science

Preparing for Your Eighth-Grade Test

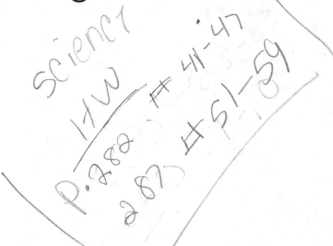

Paul S. Cohen
Former Assistant Principal, Science
Franklin Delano Roosevelt High School
Brooklyn, New York

Jerry Deutsch
Chemistry Teacher
Edward R. Murrow High School
Brooklyn, New York

Anthony V. Sorrentino, D.Ed.
Former Director of Computer Services
Former Earth Science Teacher
Monroe-Woodbury Central School District
Central Valley, New York

AMSCO

Amsco School Publications, Inc.
315 Hudson Street / New York, N.Y. 10013

The publisher wishes to acknowledge the helpful contributions of Bill Ludtka and James Delellis who acted as consultants in the preparation of this book. In addition, we wish to thank the following teachers who acted as reviewers.

Ron Benson
Educational Consultant
Former Science Teacher
Mill Middle School
Williamsville, New York

Peter Chester
Director of Science
CSD 21 K
Brooklyn, New York

Margaret Pearce
Science Editor
Amsco School Publications, Inc.
New York, New York

Anne F. Scammell
Science Teacher
Geneva Middle School
Geneva, New York

Martin Solomon
Science Teacher
Daniel Carter Beard JHS 189 Q
Queens, New York

Text and Cover Design: Howard Leiderman
Composition: Northeastern Graphic Services, Inc.
Artwork: Hadel Studio
Cartoon page 377: Courtesy of Alison Solomon

When ordering this book, please specify:
either **R 727 W** *or* REVIEWING INTERMEDIATE-LEVEL SCIENCE

Please visit our Web site at: **www.amscopub.com**

ISBN: 0-87720-183-8

To the Student

This book provides a complete review of intermediate-level science to help you prepare for your intermediate-level science test.

The text is made up of thirteen chapters. These chapters cover topics in life, physical, and earth and space science, as well as the scientific method, the history and nature of scientific inquiry, and the interactions of science, technology, and society.

The text presents the major ideas of each topic in a manner that is easy to understand. The many illustrations that go with the text help make clear the concepts presented. Each chapter is divided into several sections by topic. Each section is followed by Review Questions and Thinking and Analyzing Questions. These questions test and reinforce the main points covered in the section. The questions are designed to test your abilities in *acquiring*, *processing*, and *extending* scientific knowledge.

In addition, special features called Process Skills and Laboratory Skills appear at intervals throughout the text. The purpose of these features it to teach you a particular process-oriented or laboratory-oriented skill including reading scales, using a compass, and interpreting graphs or diagrams. Each feature guides you through the skill and then ends with several follow-up questions that have you apply the skill on your own.

Throughout the book, important vocabulary terms are printed in **bold italic type**. These terms are defined in the text and appear with formal definitions in the glossary at the back of the book. Terms that are less important or that may be unfamiliar to you are printed in *italic type*. These terms do not appear in the glossary, but are listed in the index.

Finally, the book includes three Sample Tests; the first is made up of 45 questions. The other two consist of 72 questions. These tests cover the major concepts, understandings, and skills that are included in an intermediate-level science curriculum. Again, the ability to *acquire*, *process*, and *extend* scientific knowledge is emphasized in these sample tests. We wish you success in your studies.

Contents

Chapter 1

Living Systems: Organisms

Points to Remember

▷ All living things carry out life processes. These include nutrition, respiration, transport, excretion, regulation, reproduction, and growth.

▷ All living things are composed of basic units called cells. Plant cells and animal cells have common structures. These include the nucleus, cytoplasm, and cell membrane.

▷ Cells carry out life processes.

▷ Plant cells have cell walls and chloroplasts, which are not found in animal cells. Plants manufacture their own food by a process called photosynthesis.

▷ Animals take in nutrients for energy and growth.

▷ Some microorganisms are harmful and can cause diseases called infectious diseases.

Living Things and Their Characteristics

Living Things Carry Out Life Processes

Living things, or **organisms**, share certain characteristics that set them apart from nonliving things. In particular, organisms carry out *life processes*, some of which are listed below in Table 1-1.

The Cell

Living things are made up of basic units called **cells**. Each cell of an organism carries out life processes. Certain structures are found in most cells. The **nucleus** controls cell activities. Surrounding the nucleus is a thick fluid called the **cytoplasm**, which is where most life processes occur. The cytoplasm is contained within the **cell membrane**, the "skin" of the cell, which regulates the flow of materials into and out of the cell. Figure 1-1 shows a typical animal cell.

Cells and Life Processes

Living things are made up of one or more cells. For instance, an ameba consists of a single cell, while a human consists of trillions of cells. An ameba is therefore unicellular, while a human is multicellular. Each cell carries out basic life processes. Smaller structures within the cell perform these life processes. Table 1-2 lists some of these structures.

Plant cells and animal cells have many structures in common, but they also have differences (see Figure 1-2 on page 4).

Plant cells have cell walls and cloroplast that anamol cell's dont.

1. Plant cells have a cell wall that encloses the entire cell, including the cell membrane. The tough cell wall gives support to the plant's structure. Found only in green plant cells, are structures called chloroplasts, which contain chlorophyll. Chloroplasts are the site of photosynthesis, the food-making process of plants.

Table 1-1. Life Processes and Their Functions

Process	Function
Nutrition	Taking food into the body (*ingestion*), breaking it down into a form usable by cells (*digestion*) and eliminating undigested material (*elimination*)
Transport	Moving materials throughout the organism
Respiration	Releasing energy stored in food
Excretion	Removing waste materials produced by the organism
Regulation	Responding to changes in the organism's surroundings
Reproduction	Making more organisms of the same kind
Growth	Increase in size

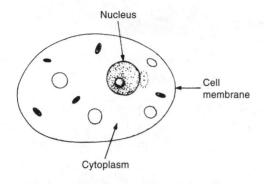

Figure 1-1. *Typical animal cell.*

2. Animal cells do not have cell walls, but are enclosed only by the cell membrane. Found only in animal cells, are the centrioles, which participate in cell division.

Organisms and Their Environment

Living things are constantly interacting with their surroundings. These surroundings are called the *environment*, which includes both living and nonliving things. Organisms obtain food, water, and oxygen from the environment. In turn, they release wastes, such as carbon dioxide. Thus, there is a continual exchange of materials between an organism and its environment.

Nutrition

Every organism needs food to stay alive. Food provides an organism with *nutrients*, which are used for producing energy as well as for growth and repair. Some important nutrients are listed in Table 1-3 on page 4. The process of *nutrition* includes ingestion, digestion and elimination. *Ingestion* is the taking in of food. *Digestion* is breaking down the nutrients into a usable form. *Elimination* or egestion is the removal of undigested materials from the body.

Table 1-2. *Some Cell Structures and Their Functions*

Structure	Function
Mitochondria	Respiration—where food is "burned" (combined with oxygen) to produce energy. Called the "powerhouse of the cell"
Ribosomes	Synthesis—where proteins are made
Lysosomes	Digestion—where digestive enzymes are stored
Nucleus	Reproduction—where genetic material is stored
Vacuole	Digestion and excretion—where digestion occurs or where excess fluid is stored
Chloroplasts	Photosynthesis—where glucose (sugar) is produced in green plants (present in plant cells only)

[handwritten margin note: were the food gets burned up energy]

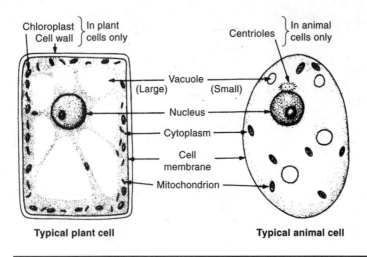

Figure 1-2. *Comparison of plant and animal cells.*

Nutrition in Plants

Green plants make their own food. Through a process called ***photosynthesis*** (Figure 1-3), plants use energy from sunlight to change carbon dioxide and water from the environment into sugar. The sun's energy is therefore stored in the sugar. Photosynthesis also produces oxygen, which is released into the environment. The green pigment ***chlorophyll***, found in the leaves and green stems of plants, is necessary for photosynthesis to take place.

Cells in the leaf of a plant contain large numbers of ***chloroplasts***, the organelles that contain chlorophyll, which is necessary for photosynthesis to take place. The roots of a plant, however, do not contain any chloroplasts. Instead, they are specialized for growing into the soil and absorbing water. The stem, with its thick cell walls, supports the plant and permits the flow of water and nutrients from the roots to the leaves and back again.

Nutrition in Animals

Animals obtain nutrients by eating plants or by eating other animals that feed on plants. The sugar in plants is used by animals to produce energy.

Table 1-3. *Nutrients and Their Uses*

Nutrient	Use
Proteins	Supply materials for growth and repair
Carbohydrates (sugars and starches)	Provide quick energy
Fats and oils	Provide stored energy
Vitamins	Assist life processes; prevent disease
Minerals	Supply materials for growth and repair; help carry out life processes

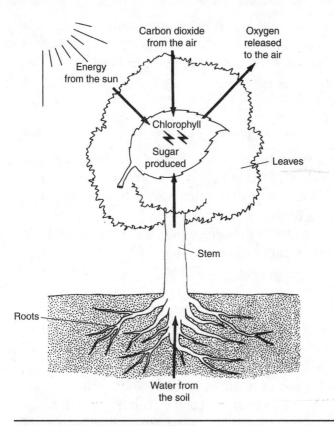

Figure 1-3. *Photosynthesis.*

The original source of energy in food is the sun. Figure 1-4 shows one way that animals get food energy from the sun.

To maintain a balanced state, all organisms must obtain a minimum daily intake of each type of nutrient. The amount of each nutrient varies based on the size, species, activity, age, and sex of the organism. An imbalance in any of these nutrients may result in weight gain, weight loss, or a disease.

Figure 1-4. *The energy in meat comes originally from the sun.*

Water is an important nutrient too. Water is constantly exchanged between an organism and its environment. For example, we get water from the environment by eating and drinking, and we give water back by exhaling, perspiring, and urinating.

Transport

Water is necessary for *transport*, the moving of materials throughout an organism. For instance, blood, which is mostly water, carries nutrients to the cells of your body. Blood takes away wastes from the cells, too. Most of the chemical processes in living things can take place only in a watery environment. In addition, water is necessary for green plants to make food. For all of these reasons, life is not possible without water.

Respiration

Organisms release the energy stored in food through a process called *respiration*, which occurs in all cells. During respiration, many nutrients in food combine with oxygen. This chemical process releases energy and forms carbon dioxide and water as waste products. Scientists use the term "burning" to describe changes in which a substance combines with oxygen and releases energy. Therefore, respiration is a form of burning without flames.

Respiration is the opposite of photosynthesis, as shown below:

Photosynthesis: energy + carbon dioxide + water → sugar + oxygen
Respiration:　　　 sugar + oxygen → energy + carbon dioxide + water

All living things, including green plants, obtain energy from food through some form of respiration.

Excretion

Carbon dioxide is a waste material produced by cells during respiration and must be removed. In humans, carbon dioxide and other waste materials are carried away by the blood. These wastes are filtered out of the blood and then removed from our bodies through exhaling, perspiring, and urinating. The process of removing wastes from the organism is called *excretion*.

Regulation

Why do you perspire when it is hot? Your body perspires to cool off. Why do you drink more when it is hot? Your body knows that it must replace water used to keep you cool. Being thirsty is a response to the heat. Organisms respond to changes in their internal and external environments. This process, called *regulation*, helps an organism maintain *homeostasis*, the maintenance of a constant internal environment.

Reproduction

Living things come from other living things. **Reproduction** is the process by which an organism produces new individuals called *offspring*. Each particular kind of organism is called a **species**. Lions are a species. Tigers are a different species. Since every individual organism eventually dies, reproduction ensures the continuation of its species.

There are two types of reproduction: asexual and sexual. **Asexual reproduction** involves only one parent. The offspring created are identical to the parent. Figure 1-5 shows examples of asexual reproduction.

Sexual reproduction involves two parents and produces offspring that are not identical with either parent. The female parent produces an egg cell, and the male parent produces a sperm cell. The joining together of these cells is called **fertilization**. The fertilized egg grows into a new individual.

Figure 1-5. Asexual reproduction: (a) binary fission; (b) budding; (c) regeneration.

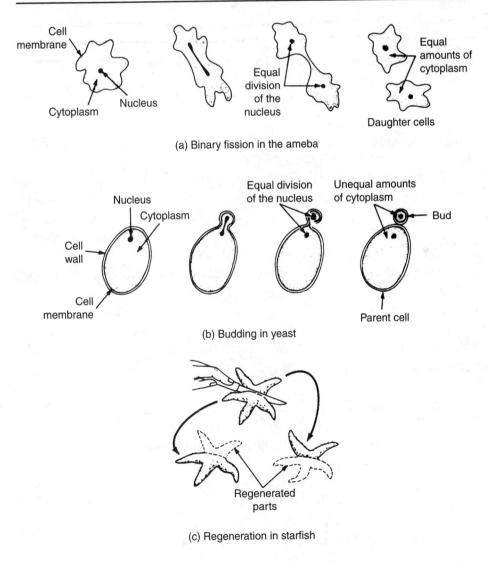

(a) Binary fission in the ameba

(b) Budding in yeast

(c) Regeneration in starfish

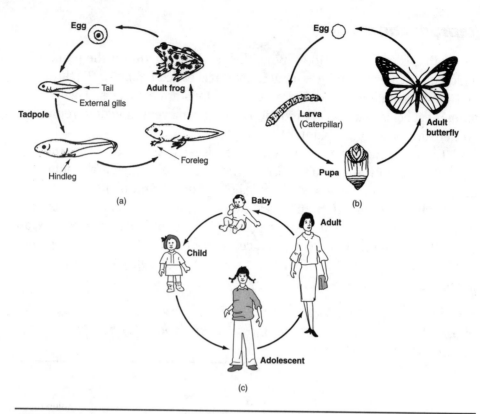

Figure 1-6. *Life cycles; (a) frog; (b) butterfly; (c) human.*

Growth

A puppy resembles an adult dog. A young elephant looks like a small version of its parents. As these young animals mature, they will increase in size. This increase in size is called *growth*. However, a frog or butterfly develops quite differently. Some organisms, such as frogs and most insects, change so dramatically during their lives that the young may not resemble the adults at all. A dramatic change such as this is called ***metamorphosis***. The changes an organism undergoes as it develops and then produces offspring make up its ***life cycle***. Figure 1-6 illustrates the life cycles of a frog, a butterfly, and a human.

 Review Questions

Multiple Choice

1. All of the following are life functions of animals *except*?

    (1) obtaining energy from food

    (2) responding to changes in the environment

    (3) production of new individuals

    (4) synthesizing food from carbon dioxide and water

2. Which organism makes its own food?

   (1) a frog   (2) a bird   (3) a tree   (4) a snake

3. The process of "burning" food inside an organism's cells to release energy is called

   (1) excretion   (2) photosynthesis   (3) digestion   (4) respiration

4. The diagram shows an example of

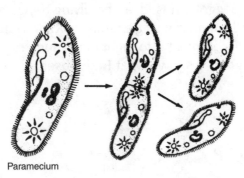

Paramecium

   (1) spontaneous generation   (3) photosynthesis

   (2) asexual reproduction   (4) respiration

5. Which statement is true of *all* living things?

   (1) They have two parents.

   (2) They are exact copies of their parents.

   (3) The young look like small adults.

   (4) They come from other living things.

Thinking and Analyzing

6. What is sugar used for by living things?

7. What is the source of energy for photosynthesis?

8. What are the two products of photosynthesis?

9. Mitochondria are also known as the "powerhouse of the cell." Which life function is carried out by the mitochondria?

10. Identify the process and describe what is happening in the diagram below.

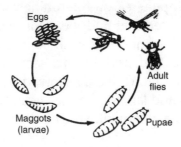

Eggs

Adult flies

Maggots (larvae)

Pupae

11. What features can be used to distinguish between an animal cell and a plant cell?

Interpreting an Experiment

About 300 years ago, the Italian scientist Francesco Redi wondered where maggots—small, wormlike organisms—come from. The popular belief at the time was that rotting meat turns into maggots. This idea, that living things could come from nonliving material, was called *spontaneous generation.* Redi designed an experiment to test this belief. He placed meat into eight jars. Four jars were left open; four were tightly sealed. Diagram 1 shows what Redi observed.

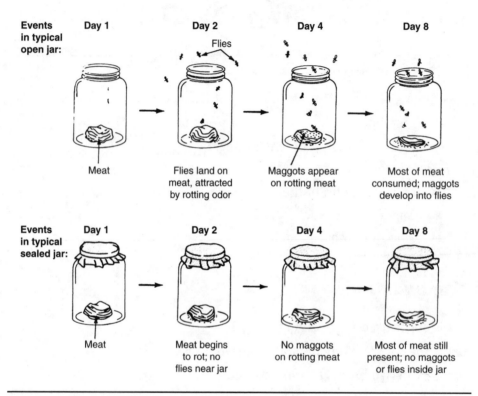

Diagram 1. Redi's first experiment: drawings show events in (top) a typical open jar; (bottom) a typical sealed jar.

As you can see, no maggots appeared on the rotting meat in the sealed jars. However, not everyone was convinced that Redi's experiment had disproved spontaneous generation. Some people claimed that fresh air was needed for spontaneous generation to occur. Therefore, Redi performed a second experiment. This time the jars were covered by fine netting, which allowed fresh air into the jars but prevented flies from entering and landing on the meat. Diagram 2 shows what Redi observed in his second experiment. Study both diagrams and then answer the following questions.

(Continued)

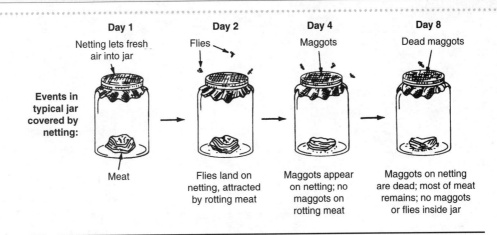

Day 1	Day 2	Day 4	Day 8

Netting lets fresh air into jar

Flies

Maggots

Dead maggots

Events in typical jar covered by netting:

Meat

Flies land on netting, attracted by rotting meat

Maggots appear on netting; no maggots on rotting meat

Maggots on netting are dead; most of meat remains; no maggots or flies inside jar

Diagram 2. Redi's second experiment.

QUESTIONS

1. Based on Redi's experiments, where do the maggots come from?
2. What conclusion can be drawn from these experiments about spontaneous generation?

Microorganisms

A *microorganism* is a very small organism that usually cannot be seen without a *microscope*. Several kinds of microorganisms are pictured in Figure 1-7 on page 12, as they would appear when seen through a microscope.

The Compound Microscope

A *microscope* is a tool used by scientists to magnify tiny objects such as cells. The *compound microscope*, used in most classrooms and laboratories, uses two lenses. They are called the *eyepiece* and the *objective*. (See Figure 1-8 on page 12) To see cells at different magnifications, most compound microscopes have more than one objective lens. The microscopes in Figure 1-8 have 10× and 40× objective lenses. By rotating the *nosepiece* we can switch from the low power objective to the high power objective. The low power objective is the shorter of the two, and has a lower magnification number. The *magnification* of a microscope can be calculated by multiplying the eyepiece power by the objective power. For example, the eyepiece in the microscopes in Figure 1-8 is 10×. Since the low power objective lens is also 10×, the low power magnification must be 100×. What would be the high power magnification? The eyepiece is 10× and the high

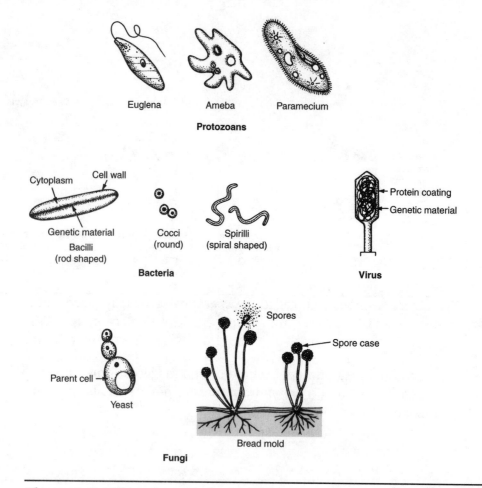

Euglena Ameba Paramecium

Protozoans

Cytoplasm Cell wall

Genetic material

Bacilli
(rod shaped)

Cocci
(round)

Spirilli
(spiral shaped)

Bacteria

Protein coating
Genetic material

Virus

Spores

Spore case

Parent cell

Yeast

Bread mold

Fungi

Figure 1-7. *Different types of microorganisms.*

Figure 1-8. *Compound microscopes.*

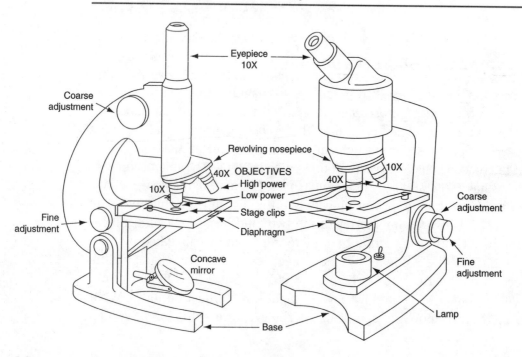

Eyepiece
10X

Coarse
adjustment

Revolving nosepiece

40X OBJECTIVES
High power
Low power

10X

Stage clips

Diaphragm

Fine
adjustment

Concave
mirror

10X

40X

10X

Coarse
adjustment

Fine
adjustment

Lamp

Base

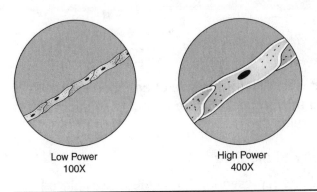

Low Power
100X

High Power
400X

Figure 1-9. *More cells are seen under low power than under high power.*

power objective is 40×. We multiply 10 × 40 and get 400. The magnification under high power is 400×.

The object to be viewed, called the **specimen**, is usually placed on a glass *slide*. The slide is placed on the *stage* and kept in place by the *clips*. A mirror or lightbulb is used to project light through the opening in the stage. The light passes through the specimen, through the objective lens, through the eyepiece, and finally to the eye.

The microscope has two knobs that are used to focus the image. The **coarse-adjustment knob** is always used first under low power. Once the image is in focus under low power, it is possible to switch to the high power objective. All additional focusing adjustments must be done with the **fine-adjustment knob**. This prevents the longer high power objective lens from hitting the slide and possibly breaking.

When viewing the image, you will notice that it appears upside down, and moves in the opposite direction from the movement of the slide. The area you see is called the **field of view**. The field of view is greater under low power than it is under high power. For example, if you were viewing a group of cells, (see Figure 1-9) you would see a larger number of cells under low power than you would under high power. If the magnification were four times greater under high power, the diameter of the field would be one fourth of what it is under low power. However, most microorganisms are so small that they must be viewed under high power.

Laboratory Skill

Estimating the Size of a Cell

A microscope can be used to view things that are very small. Microscopes are used to look at cells. To clearly see the parts of the cell, a material called a **stain** is used to color certain of the organelles. In this experiment, Juan was asked to stain an onionskin and estimate the size of the cells.

(Continued)

The problem for this experiment was: What is the size of an onion skin cell?

His materials were: a compound microscope, tincture of iodine solution, glass slide, cover slip, dropper, clear plastic ruler, paper towel.

Procedure 1—Estimating the diameter of the field

- A clear plastic ruler with each small line representing 1 millimeter was placed on the microscope stage. See Diagram 1.

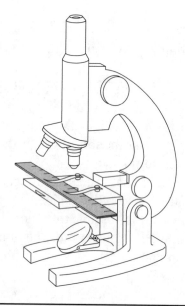

Diagram 1. *Placing the ruler on the stage.*

- It was observed under low power.

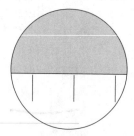

Diagram 2. *Measuring the diameter of a field under low power.*

QUESTIONS

1. What is the approximate diameter of the field under low power?
2. The microscope used provides a magnification of 100× under low power, and 400× under high power. What is the approximate diameter of the field under high power?

(Continued)

Reviewing Intermediate-Level Science

Procedure 2—Observing cells under low power

- An onionskin was placed in the center of a glass slide. A drop of water was placed on it, and it was then covered with a coverslip as shown in Diagram 3.

Diagram 3. *Covering the specimen with a coverslip.*

- The onionskin was then stained with tincture of iodine as shown in Diagram 4.

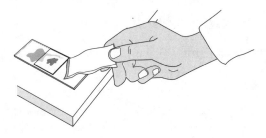

Diagram 4. *"Drawing" a stain across a slide.*

- When viewed under low power, the cells appeared as shown in Diagram 5. Count the number of cells that fit across the diameter of the field of view.

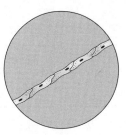

Diagram 5. *Viewing onion cells under low power.*

- By dividing the diameter of the field (found in Part 1) by the number of cells that fit across the field of view, Juan could determine the average size of each cell.

(Continued)

3. What was the average size of each cell?
4. What was the purpose of staining the cells with tincture of iodine?

Procedure 3–Observing under high power

- Juan was asked to predict how many cells he would see under high power and then check his prediction. Recall that the microscope magnifies 100× at low power and 400× at high power.

QUESTIONS

5. How many cells could Juan see under high power?
6. Which part of the microscope did Juan use to focus the cells under high power?
7. If Juan measured the average size of the cells under high power, how would his result compare with the value he obtained at low power?

Harmful Microorganisms

Certain microorganisms can be harmful to humans and other living things. These are often called germs. An *infectious disease* is an illness caused by microorganisms that can be transmitted, or passed on, from one individual to another. Table 1-4 lists some infectious diseases and the types of microorganisms that cause them.

Noninfectious Diseases

Not all diseases are caused by microorganisms. Arthritis, high blood pressure, asthma, and cancer are examples of *noninfectious diseases*, which

Table 1-4. Some Infectious Diseases and Their Causes

Disease	Type of Microorganism
Pneumonia	Bacteria
Strep throat	Bacteria
Botulism	Bacteria
Common cold	Viruses
Flu (influenza)	Viruses
Mono (infectious mononucleosis)	Viruses
AIDS	Viruses
Athlete's foot	Fungus
Amebic dysentery	Protozoan

Table 1-5. Some Noninfectious Diseases and Their Causes

Disease	Cause
Scurvy	Vitamin C deficiency
Anemia	Iron deficiency
Hay fever	Allergy to pollen
Gastric ulcer	Excess stomach acid
Hemophilia	Heredity
Sickle Cell Anemia	Heredity
Diabetes	Malfunctioning pancreas

cannot be transmitted from one individual to another. Causes of noninfectious diseases include poor diet, malfunctioning glands, damaged organs, allergies to foreign substances, and reactions to chemicals or radiation in the environment. There are also diseases that are hereditary; that is, they are inherited from one's parents. These diseases can only be transmitted from parent to child. Table 1-5 lists some noninfectious diseases and their causes.

Cancer is the result of abnormal cell division. Scientists are not sure exactly what causes this growth. They do know of certain environmental factors that increase the occurrence of cancerous cells. For example, cigarette smoking is known to increase the probability of getting lung cancer. Excessive exposure to sunlight may cause skin cancer.

Helpful Microorganisms

Only about 5 percent of the known microorganisms are harmful. In fact, many microorganisms are beneficial to us and to other living things, and some are even essential to our well-being. Helpful microorganisms include the *bacteria of decay*, or other *decomposers*, which break down dead organisms and return nutrients to the environment.

Some microorganisms are used to help produce certain foods. Yogurt, for example, is produced by the action of bacteria on milk. *Molds,* a type of fungus, help in making cheese. *Yeasts* are fungi that cause bread to rise and produce alcohol in beer and wine. Humans even need certain kinds of bacteria inside their bodies to aid in digestion.

Experimental Design

In the late nineteenth century, the French scientist Louis Pasteur investigated a disease called *anthrax,* which was killing sheep and cattle. He suspected that the disease was caused by a particular kind of bacteria. To test this *hypothesis,* or educated guess, he performed an experiment.

Pasteur heated a sample of the bacteria just enough to weaken, but not kill, them. Using a group of 50 sheep, he injected 25 sheep with the weakened bacteria, and left the other 25 sheep alone. The injected sheep became slightly ill, but soon recovered. Several weeks later, Pasteur injected all 50 sheep with a large dose of healthy bacteria, strong enough to kill a normal sheep. After a few days, all 25 sheep that had been injected with weakened bacteria were still alive, while the other 25 were all dead of anthrax.

Through this experiment, Pasteur demonstrated that a type of bacteria in fact caused anthrax. He also showed that by giving sheep a mild case of the disease, he could protect them from more serious infection in the future. This procedure is called *immunization,* and is used today to protect humans from many infectious diseases.

Pasteur's demonstration was effective because he followed the scientific method in designing his experiment. Why was it necessary to use two groups of sheep? The group that was made immune by the first injection was the *experimental group.* To be sure it was the injection that made them immune, another group was needed for comparison. This was the *control group.* Both groups had to be treated exactly the same, except for the condition that was being tested. The condition that was different (the immunization) was the *independent variable.*

In an experiment, the condition that is manipulated, or changed, by the scientist is called the independent variable. The condition that responds to the change is called the *dependent variable.* In Pasteur's experiment, the dependent variable was the sheep's immunity to anthrax.

A teacher asked her students to perform an experiment on factors that affect the souring of milk. Betty obtained three containers of "Surefresh" milk. One she kept at room temperature, 20°C. Another container she refrigerated at 5°C, and the third she kept at 1°C. She checked the milk every day. Examine her results in the table below, and answer the questions that follow.

Temperature	Time Until Milk Turned Sour
20°C	1 day
5°C	3 days
1°C	7 days

(Continued)

QUESTIONS

1. What was the independent variable in this experiment?
2. Betty might reasonably conclude that as the temperature increases, the length of time that milk takes to turn sour
   (1) increases   (2) decreases   (3) remains the same
3. Based on her experiment, Betty should predict that at 10°C, the milk would turn sour in about _____ days.

Marshall decided to compare the souring times of three different brands of milk. He left a container of "Surefresh" milk at 20°C, one of "Sunshine" milk at 5°C, and one of "Dairytime" milk at 1°C. All three containers had the same expiration date stamped on them, and all were fresh when the experiment began. Here are his results:

Brand	Temperature	Time to Sour
Surefresh	20°C	1 day
Sunshine	5°C	2 days
Dairytime	11°C	8 days

QUESTIONS

4. Which conclusion could Marshall reasonably draw from his experiment?
   (1) "Surefresh" milk turns sour faster than "Sunshine."
   (2) "Sunshine" milk turns sour faster than "Dairytime."
   (3) All brands of milk are the same.
   (4) He couldn't conclude anything from the experiment.
5. What was the major mistake that Marshall made in his experiment?

Marshall's teacher told him to check all three brands at the same temperature. When he did this, at 5°C, he got the following results:

Brand	Time to Sour
Surefresh	3 days
Sunshine	2 days
Dairytime	2 days

All three containers had the same expiration date and were fresh before the experiment began. Marshall concluded that he should always buy "Surefresh." Other students, however, repeated his experiment and found that, on the average, there was no difference among the three brands.

(Continued)

6. What was wrong with Marshall's conclusion?
(1) He didn't measure the time accurately enough.
(2) 20°C is too warm to store milk.
(3) He should have tried several different temperatures.
(4) He should have repeated the experiment several times to see if he always got the same results.

Review Questions

Multiple Choice

12. Infectious diseases are caused by
(1) microorganisms
(2) allergies
(3) poor diet
(4) chemicals in the environment

13. Which disease is a type that can be transmitted from one individual to another?
(1) hay fever (2) the flu (3) skin cancer (4) diabetes

14. Noninfectious diseases can be caused by all of the following *except*
(1) malfunctioning glands   (3) bacteria
(2) damaged organs   (4) reaction to chemicals in the environment

15. Scurvy is a disease that sailors often used to get on long voyages. It was found that eating oranges and limes prevented scurvy. This suggests that scurvy is a disease caused by
(1) a microorganism   (3) an allergy
(2) a deficiency in diet   (4) a damaged organ

16. Studies have determined that smoking cigarettes can cause lung cancer. This is an example of a disease caused by
(1) reaction to chemicals in the environment
(2) microorganisms
(3) deficiencies in diet
(4) bacteria

Questions 17 and 18 refer to the diagram below that represents what is seen through a microscope when looking at some cells.

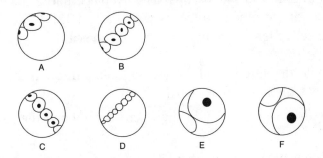

17. A student wished to examine some cells in a microscope. When she first focused on the slide at low power, the cells appeared as shown in diagram A above. To center the cells, as shown in diagram B, she should move her slide

    (1) down and to the left   (3) up and to the left
    (2) down and to the right   (4) up and to the right

18. After centering the slide, she switched to high power. To which diagram would the cells now appear most similar?

    (1) C   (2) D   (3) E   (4) F

Thinking and Analyzing

19. Lyme disease is an infectious disease transmitted to humans by a deer tick (a small animal related to spiders). The numbers in the diagram below show how many cases of Lyme disease were reported in 1999 in several northeastern states. Make a list of all the states indicated in order of the number of reported cases of Lyme disease starting with the most number of reported cases and ending with the least. Which state had the second-most reported cases of Lyme disease?

Northeastern U.S.

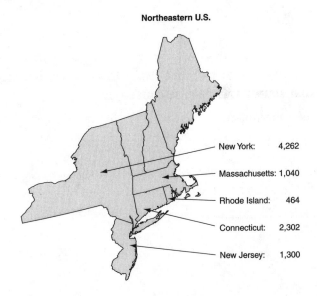

New York: 4,262

Massachusetts: 1,040

Rhode Island: 464

Connecticut: 2,302

New Jersey: 1,300

Read the following paragraph to answer questions 20–22.

A student found an old microscope. The eyepiece had a magnification of 10×, and one of the objective lenses was marked 12×. The markings on the other objective lens have been rubbed off over time.

20. How could the student tell, before he looked through the microscope, which was the high power lens?

21. What is the magnification of the microscope when the 12× objective is used?

22. The student measures the diameter of the field, using the 12× objective, and finds it to be 1.6 millimeters. Using the other, unmarked objective, the diameter of the field is 0.40 millimeters. What is the magnification of the microscope using the unmarked objective lens?

23. Below you will find the steps followed to correctly use a compound microscope. However, these steps are not in the proper order. Arrange these steps in the proper order.

(a) Use the fine focus to see the object clearly.

(b) Adjust the mirror or light source.

(c) View the object under low power.

(d) Use the coarse focus.

(e) View the object under high power.

Chapter 2
Living Systems:
Ecosystems

Points to Remember

▷ Living things and the nonliving factors in their surroundings make up an ecosystem. Adaptations allow an organism to survive in a particular type of ecosystem.

▷ Ecological succession is a natural process by which one community is replaced by another community in an orderly, predictable sequence.

▷ Green plants use sunlight to make their own food. They are called producers. Animals depend on other organisms for food. They are called consumers. All organisms get their energy directly or indirectly from the sun.

▷ Decomposers are organisms that break down the remains of dead plants and animals. Decomposers return nutrients to the environment.

▷ Producers, consumers, and decomposers may be linked in a sequence called a food chain. Disturbing any part of a food chain affects other organisms in the food chain.

▷ There is a continual exchange of materials between an organism and its environment: these materials include food, water, oxygen, and wastes.

Living Things and Their Environment

Stimulus and Response

Living things must react to changes in their environment. For example, when the air gets too hot, you perspire. When the light gets too bright, the pupils of your eyes get smaller. Changes in the environment are called *stimuli*. The ways in which living things react to these changes are called *responses*.

Sometimes an environment is harsh, which makes it difficult for an organism to survive. Living things have evolved in response to their environments by developing *adaptations*, characteristics that help them survive in their habitat.

Behavioral Adaptations

Some environments undergo extreme changes in temperature, amount of sunlight, and water supply from season to season. Organisms that live in such environments have special behaviors called *behavioral adaptations* that help them adapt to these changes. These include migration, hibernation, and dormancy.

1. *Migration.* Many birds that live in cold climates fly to warmer regions as winter approaches. This is called *migration*, moving from one environment to another.
2. *Hibernation.* Some animals, such as bears, survive the cold by sleeping for most of the winter. This is called *hibernation*.
3. *Dormancy.* Other living things may adjust to extreme environmental changes by entering a state of *dormancy*, becoming completely inactive. During the winter when a tree has lost its leaves, the tree is not dead—it is *dormant*. When spring comes, bringing warmer conditions, the tree grows new leaves. A seed may remain dormant for years, waiting for the proper conditions for growth.

Some organisms survive for only one season. Plants, called annual plants, go through their entire life cycle, including reproduction, during that season. Some insects such as the mosquito also live for only one season. Their larvae survive the winter and hatch the next summer.

Physical Adaptations

Living things have developed special characteristics called *physical adaptations* that enable them to survive under a given set of conditions. Organisms may be adapted for life in water, soil, or air. For example, a fish has gills so it can take in oxygen from the water. An earthworm's body shape helps it move through the soil. A bird has wings and light, hollow bones so it can fly. Many adaptations help an organism to obtain food or

Figure 2-1. *Adaptations of birds to their environment.*

escape predators in its environment. Figure 2-1 shows how certain birds are adapted for survival.

Earth's many environments include oceans, deserts, tropical rain forests, and the frozen Arctic tundra. Adaptations permit an organism to live in its own particular environment, or **habitat**. Organisms living in a dry, desert environment have adaptations that enable them to obtain and conserve water. For example, the cactus plant has an extensive root system that helps it reach what little water there is in the desert.

Animals living in the icy Arctic have adaptations that help them to endure the region's very cold temperatures. For instance, polar bears have thick coats of fur; seals and whales have layers of protective fat called *blubber*. Table 2-1 lists organisms from various habitats and their adaptations.

Table 2-1. *Organisms and Their Adaptations*

Organism	*Habitat*	*Adaptation*	*Function*
Giraffe	Grasslands	Long neck	Helps to reach leaves on trees, its main food
Arctic Hare	Arctic	White fur in winter	Provides camouflage from predators
Monkey	Rain forest	Grasping tail	Acts as an extra hand, freeing hands and feet for other uses
Cactus	Desert	Waxy skin	Reduces water loss from evaporation

Extinction

An organism will survive only if it can adjust to changes in its environment. A species that cannot adapt to major changes in its environment will become *extinct*. Extinction occurs when a species dies out. This usually happens when some essential part of the living or nonliving environment is removed. Common natural causes of extinction include natural disasters, climate changes, and habitat invasion by a predator. Humans also cause extinction by excessive hunting and pollution of land, air, and water.

Scientists know of thousands of species that once existed but are now extinct. Evidence of these extinct organisms can be found in fossils. *Fossils* are the remains or traces of organisms that have lived in the past. Fossils include skeletons, shells, and impressions, such as footprints, preserved in rock. Some animals that have become extinct are dinosaurs, woolly mammoths, dodos, and passenger pigeons.

About 65 million years ago, roughly 70 percent of all animal species, including the dinosaurs, became extinct. Scientists are not sure what caused this mass extinction. Some have suggested that there was a sudden change in climate. Perhaps an asteroid or comet collided with Earth causing huge clouds of dust that blocked the sun. Only those species that were able to adapt to these drastic changes in the environment survived. The dinosaurs could not adapt successfully and died out. The elimination of the dinosaurs helped mammals to survive, evolve, and eventually become the dominant large animals that they are today.

Many species have become extinct due to the direct interference of human beings. In 1598, Portuguese sailors discovered the dodo, a large, flightless bird on an island in the Pacific Ocean. (See Figure 2-2.) These

Figure 2-2. *The dodo (left) is extinct; the bald eagle is threatened.*

birds had no natural enemies, so they did not fear the sailors, and were easily killed. Those that survived were killed by the dogs and pigs introduced to the island by the sailors. By 1681, the dodo was completely eliminated.

Learning from the mistakes of the past, we now have laws protecting species that are close to extinction. Many of these ***endangered species***, such as the California condor and Florida panther, have been rescued from extinction. The American bald eagle was once believed in danger of extinction, but due to the efforts of conservationists the eagle is no longer endangered. However, it is still considered a threatened species. (See Figure 2-2.)

Communities and Ecosystems

A habitat usually contains many different types of organisms that interact with one another and may depend on each other for survival. Within the habitat, all the members of a particular species are called a ***population***. All the different populations within a habitat make up a ***community***. When you set up an aquarium containing plants, catfish, and guppies, you create a small community.

To set up an aquarium, you must provide more than just the fish and the plants. You need water, a source of oxygen, and light. You must also maintain a proper temperature. These nonliving factors together with the living members of the community make up an ***ecosystem***.

The members of the community get the materials they need to survive from the ecosystem. In return, they give materials back, such as wastes and dead, decaying bodies. Materials are constantly being recycled within an ecosystem. Figure 2-3 shows how oxygen and carbon dioxide are recycled. Most of the oxygen in our environment is provided by green plants through the process of photosynthesis. Energy, however, is not recycled and must be provided by an outside source, such as the sun.

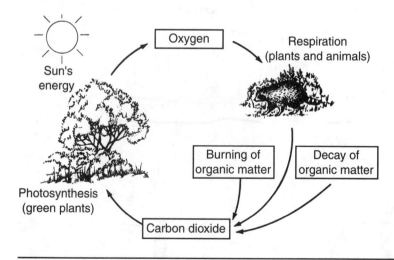

Figure 2-3. *Oxygen and carbon dioxide are constantly recycled in an ecosystem.*

Process Skill

Interpreting the Results of an Experiment

A student performed an experiment in which five plants were placed in sand and five plants were placed in soil. All 10 plants were the same type, given an equal amount of water, and exposed to an equal amount of sunlight. The experiment lasted for two weeks. The table below shows the growth of each of the plants.

Plants in Soil	Increase in Height (in centimeters)	Plants in Sand	Increase in Height (in centimeters)
1	2.0	1	0.5
2	1.9	2	0.6
3	2.2	3	0.4
4	2.1	4	0.7
5	1.9	5	0.6

QUESTIONS

1. What conclusion may be drawn from this experiment?
   (1) Plants grow just as well in soil as in sand.
   (2) Plants grow taller in sand than in soil.
   (3) Plants grow taller in soil than in sand.
2. What differences between the soil and the sand would explain the results of this experiment?
3. Which bar graph correctly represents the averaged results of this experiment?

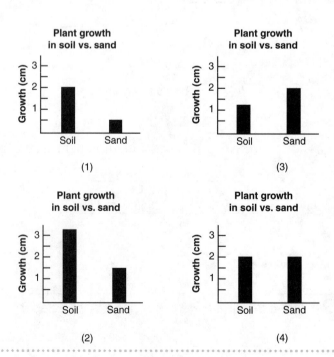

Multiple Choice

1. A change in the environment is called a

   (1) stimulus (2) response (3) migration (4) reaction

2. A student touches a hot object and quickly pulls his hand away. This is an example of

   (1) a response followed by a stimulus

   (2) a stimulus followed by a response

   (3) a response followed by a response

   (4) a stimulus followed by a stimulus

3. Migration, hibernation, and dormancy are all methods of

   (1) producing food

   (2) changing the environment

   (3) adjusting to changes in the environment

   (4) producing energy

4. In which environment would you most likely find an animal with thick fur?

   (1) desert (2) tropical rain forest (3) Arctic (4) grassland

5. Life in the desert is difficult because there is very little

   (1) sunshine (2) water (3) oxygen (4) sand

6. All of the organisms that live in a pond make up

   (1) a habitat   (3) an environment

   (2) a community   (4) an ecosystem

Thinking and Analyzing

7. Birds that are adapted to live in a watery environment would most likely have the type of feet shown in

(1)

(3)

(2)

(4)

Questions 8 through 11 refer to the diagram below, which represents a small ecosystem.

8. The survival of this community depends upon a constant external supply of

   (1) energy   (2) oxygen   (3) carbon dioxide   (4) plants

9. What is the main source of oxygen in this ecosystem?

10. Make a list of the living and nonliving components of this ecosystem.

11. If light is eliminated from this environment, the fish die. Explain why.

The Balance of Nature

Producers and Consumers

All organisms need energy to survive. They get this energy from nutrients in food. During photosynthesis, green plants produce sugar, our main source of food energy. Plant-eating animals, called **herbivores**, obtain this energy-rich sugar when they eat and digest the plants.

Meat-eating animals, called **carnivores**, also get energy from plants, but indirectly. For instance when a lion eats a zebra, it obtains nutrients from the meat of the zebra. The lion gets its energy from these nutrients. The zebra had obtained these nutrients from the plants that are its food.

Both the lion and the zebra depend upon other organisms for their food. Green plants produce their own food. Therefore, green plants are called **producers**, while animals are called **consumers**. **Omnivores** are consumers too. Omnivores are animals that can eat either plants or other animals. Every animal depends directly or indirectly on green plants for food and oxygen.

Food Chains

The nutrients in green plants get passed along from one organism to another in a sequence called a **food chain**. Grass produces food during photosynthesis. A zebra eats the grass to get its nutrients. A lion, in turn, eats the zebra.

When the lion dies, its body decays. Special organisms called **decomposers** break down the lion's remains and return nutrients to the soil. Plants, such as grass, can then use these nutrients again.

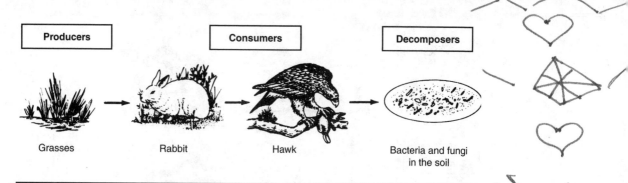

| Producers | Consumers | Decomposers |

Grasses → Rabbit → Hawk → Bacteria and fungi in the soil

Figure 2-4. A food chain.

Decomposers include fungi, such as mushrooms and molds, and some bacteria. Fungi and bacteria cannot make their own food, and so depend on other living things for food. Decomposers are the last link in any food chain. Figure 2-4 shows an example of a food chain.

Suppose that, suddenly, there were no more zebras. How would the lion get its nutrients? A lion can also eat other animals. If something wiped out the zebra population, the lion would eat more of the other animals. Thus, the removal of one species, the zebra, would affect many other species.

Food Webs

Most ecosystems contain a number of food chains that are interconnected to form a *food web*, as shown in Figure 2-5 on page 32. There is a delicate balance in an ecosystem among its producers, consumers, decomposers, and their environment. If this balance is disturbed, it may change the entire ecosystem.

Food Pyramid

What happens to you when you exercise? Exercising causes you to "burn" a lot of food for energy. As you do this, your body temperature increases. Heat is being generated, and is lost to the environment. Every organism uses some of the energy it consumes and stores the rest. The energy that is used is lost to the environment in the form of heat. Only the stored energy is available to the next consumer in the food chain. This results in a decreasing amount of energy available at each step of the food chain. This can be represented by a pyramid, which gets smaller and smaller toward the top. Figure 2-6 on page 32 illustrates an energy, or food pyramid. What type of organism is always at the base? A food pyramid always has green plants (producers) at its base.

Symbiosis

When different organisms live together, they may interact in a variety of ways. At least one organism always benefits from the relationship. The other organism, however, may or may not benefit. **Symbiosis**, or a *symbiotic relationship*, occurs when two or more different organisms live in close association with one another; that is, when one organism lives on or inside another

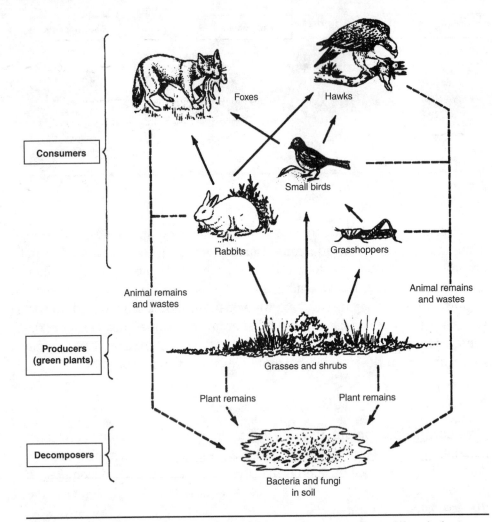

Figure 2-5. A food web consists of several interconnected food chains.

Figure 2-6. A food pyramid shows the relationship of producers to consumers; the producers (plants) are always at its base.

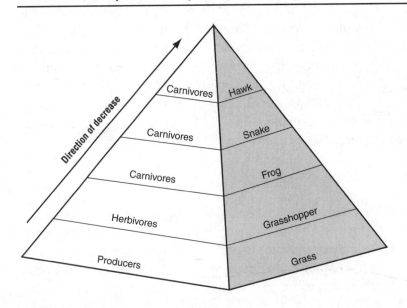

Table 2-2. Types of Symbiotic Relationships Between Organisms

Relationship	Description
Mutualism	Both organisms benefit from the relationship.
Commensalism	One organism benefits while the other is not affected.
Parasitism	One organism benefits while the other organism is harmed.

one. For example, a flea living on the skin of a dog has a symbiotic relationship with the dog; this type of symbiosis is called parasitism because the flea benefits while the dog is harmed. On the other hand, bacteria living inside termites enable the termites to digest wood. The bacteria as well as the termites benefit from this arrangement. This relationship is called mutualism. In commensalism, another symbiotic relationship, one organism benefits while the other is not affected. Orchids and the trees on which they grow have a commensal relationship. The tree only supplies support. The roots of the orchids absorb what they need from the air. Table 2-2 describes the three types of symbiotic relationships that may exist between organisms in a community.

Competition

Competition for food and space is an important part of the relationships among living things in an ecosystem. Both the moose and the snow shoe hare live in the same habitat and compete for food from the birch tree. The moose, by far the larger of the two animals, has an advantage over the hare in obtaining food. In winter, when food is scarce, the hare is more likely to die of starvation because of competition for the limited food available. Plants also compete for space and sunlight. (See Figure 2-7.)

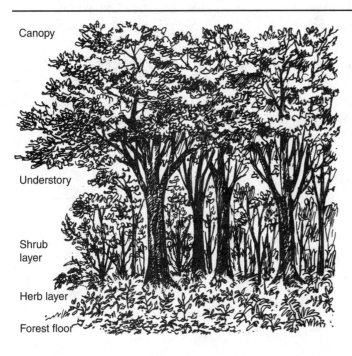

Canopy

Understory

Shrub layer

Herb layer

Forest floor

Figure 2-7. Plants compete for resources, such as growing space and sunlight.

Upsetting the Ecosystem

Biodiversity describes the variety of life-forms that exist. There is great value to having a large number of species living on Earth. Humans should preserve biodiversity because we depend on plants and animals for such things as food, medicine, and much of our industry.

Humans sometimes interfere with the balance of nature. For example, the early settlers in the northeastern United States killed off all the wolves in the region because the wolves preyed on farm animals. However, the wolves were the only natural enemies of the deer living in the area. Without the wolves to hold their numbers down, the deer population increased to the point where many deer starved to death in winter.

The actions of people are not the only things that can disturb the balance of nature. (See Table 2-3.) Sometimes, the delicate balance may be upset suddenly by natural events such as floods and forest fires. In 1980, for instance, a volcano called Mount St. Helens, in the state of Washington, erupted violently. The explosion destroyed almost 100,000 acres of forest. With time, the forest is returning to Mount St. Helens through a series of natural changes in the ecosystem. In the summer of 2000, many areas in the Northwest and Texas were badly damaged by forest fires. In time, these areas also will regrow.

Ecological Succession

After a forest fire or volcanic eruption has destroyed an ecosystem, the soil becomes enriched with minerals from the decaying remains of the plants and animals that had lived there. Soon, small new plants sprout. These become homes and food for insects and small animals. Eventually, these plants die, and are replaced by other, larger plants. Each new community changes the environment, making it more suitable for the next community. Finally, a community emerges that is not replaced, called the climax community.

On Mount St. Helens, the climax community is the forest of spruce and fir trees along with the animals that lived there, which existed before the eruption. The natural process by which one community is replaced by another in an orderly, predictable sequence is called *ecological succes-*

Table 2-3. *Causes of Change in Ecosystems*

Gradual Changes	Sudden Change
Succession	Volcanic eruptions
Climate change	Forest fires
Human population growth	Human actions
Continental drift	Floods, meteorite impacts

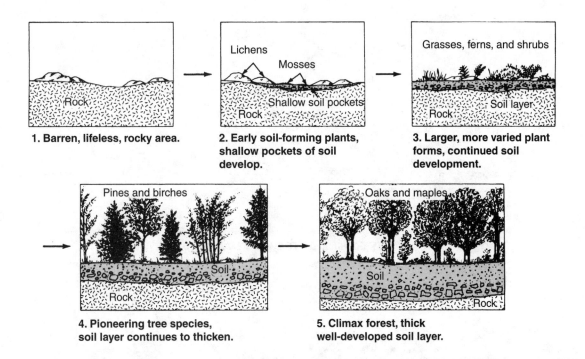

| 1. Barren, lifeless, rocky area. | 2. Early soil-forming plants, shallow pockets of soil develop. | 3. Larger, more varied plant forms, continued soil development. |

| 4. Pioneering tree species, soil layer continues to thicken. | 5. Climax forest, thick well-developed soil layer. |

Figure 2-8. Ecological succession on land.

sion. Figure 2-8 illustrates the ecological succession of a barren area into a forest.

Conserving Natural Resources

A forest is an important *natural resource*. It supplies wood and oxygen, conserves soil and water, and provides a habitat for wildlife and recreation for people.

Forests that are destroyed can be replaced, although renewal takes a long time. This means the forest is a **renewable resource**, a resource that can be replenished. When plants and animals die and decay, their nutrients are returned to the soil. Therefore, soil is a renewable resource. Water, too, is a renewable resource, since it is constantly recycled through the environment (Figure 2-9 on page 36).

Aluminum, like other minerals, is not replenished by nature. Minerals and other materials that are not naturally replaced are **nonrenewable resources**. To guarantee an adequate supply of these valuable materials in the future, we must conserve and recycle them today.

Although nature does recycle water, soil, and forests, humans often use them up faster than nature can replace them. It is important, therefore, to conserve these resources as well, or we may have shortages of them someday.

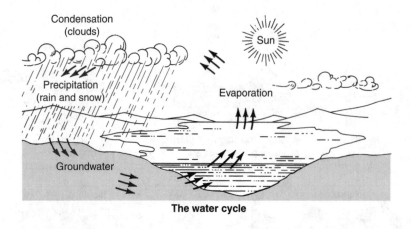

Figure 2-9. *Water is constantly recycled though the environment.*

  Review Questions

Multiple Choice

To answer questions 12 to 14, use the following diagram, which represents a food chain.

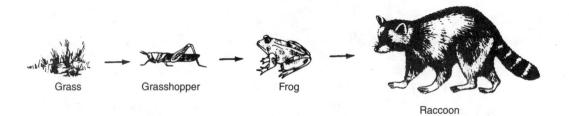

12. Which organism is the producer in this food chain?

    (1) grass   (3) frog

    (2) grasshopper   (4) raccoon

13. In this food chain, the frog is a

    (1) producer

    (2) consumer

    (3) decomposer

14. Which type of organism is not shown in this diagram?

    (1) producer

    (2) consumer

    (3) decomposer

To answer questions 15 to 18, use the diagram below of a food web.

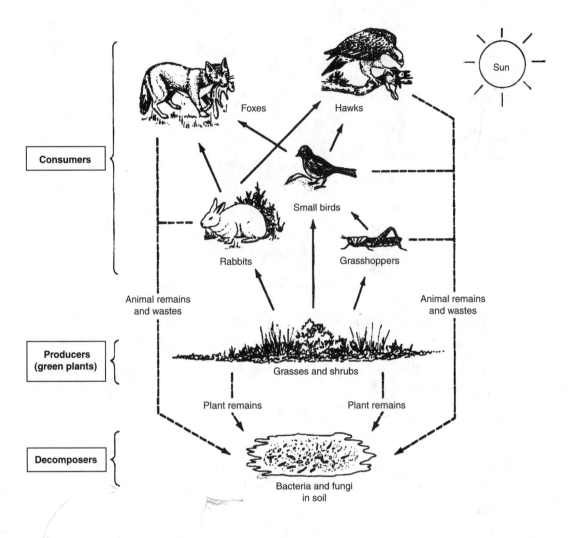

15. A decrease in the number of small birds would most likely result in an increase in the number of

(1) rabbits   (2) hawks   (3) foxes   (4) grasshoppers

16. The organisms that return nutrients to the soil are the

(1) producers   (3) decomposers

(2) consumers   (4) green plants

17. The producers in this ecosystem get their energy from

(1) the rabbit   (3) the decomposers

(2) the consumers   (4) the sun

18. Animals that eat only plants are called herbivores. Animals that eat only meat are called carnivores. Animals that can eat both meat and plants are called omnivores. According to the food web, which organism may be considered an omnivore?

(1) grasshopper   (2) small birds   (3) hawk   (4) fox

Base your answer to question 19 on the diagram below, which represents a food pyramid.

19. Which organism would be considered a producer?

    (1) the human, because it is at the top of the pyramid

    (2) the pig, because it changes the energy in plants into food for the human

    (3) the corn, because it changes energy from the sun into food

    (4) the corn, because it cannot be eaten by the human

20. As the population of old shrubs decreases in a changing ecosystem, the population of new trees increases. The old community

    (1) destroys the ecosystem

    (2) prepares the ecosystem for the new community

    (3) is the climax community

    (4) does not provide nutrients to the soil

21. Which organism makes its own food?

    (1) frog   (2) bird   (3) tree   (4) snake

22. Photosynthesis is the process by which plants make their own food. The energy for photosynthesis comes from

    (1) the sun   (2) oxygen   (3) water   (4) wind

23. All of the organisms that live in a pond make up

    (1) a habitat   (3) the environment

    (2) a community   (4) an ecosystem

Base your answers to questions 24 and 25 on the following diagram, which represents a small ecosystem.

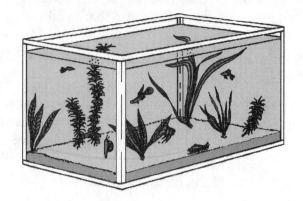

24. All animals need oxygen to live, even animals that live in the water. What is the main source of oxygen in this ecosystem?

(1) the water   (2) the fish   (3) the snail   (4) the green plants

25. Many important, nonliving factors within an ecosystem are recycled, although some are not. The survival of this community depends upon a constant external supply of

(1) energy   (2) oxygen   (3) carbon dioxide   (4) plants

Base your answers to questions 26 and 27 on the diagram below, which represents a food pyramid.

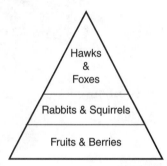

26. Based on this food pyramid, which would provide the greatest quantity of energy per pound when eaten?

(1) apples   (2) rabbits   (3) hawks   (4) foxes

27. Which organism can be considered a carnivore?

(1) rabbits, because they eat berries

(2) berries, because they produce sugar from the sun's energy

(3) squirrels, because they eat acorns

(4) hawks, because they eat animals

28. Which symbiotic relationship is an example of parasitism?

(1) Bacteria live in our intestines and help us digest our food.

(2) Ticks live on a dog's skin and suck its blood and transmit diseases.

(3) Remora fish attach themselves to a shark and eat its leftover food bits.

(4) Barnacles attach themselves to whales and are transported to new feeding grounds.

29. Certain crocodiles allow a small bird called an Egyptian plover to sit inside their open mouth. The birds feed on harmful leeches and food particles found between the teeth of the crocodile. This relationship is best described as a type of

   (1) parasitism, because the crocodile is harmed

   (2) mutualism, because both organisms benefit

   (3) commensalism, because the crocodile is not affected

   (4) parasitism, because the bird is harmed

30. Small fish called clownfish are immune to the stinging cells of a sea anemone's tentacles. The clownfish live within these tentacles, where they are protected from larger fish (see figure, below). The sea anemone also benefits, because it can feed on scraps of food left over by the clownfish. This type of relationship is best described as

   (1) parasitism

   (2) mutualism

   (3) communalism,

   (4) a food web

31. Which of the following is a nonrenewable resource?

   (1) soil   (2) silver   (3) water   (4) forest

Thinking and Analyzing

32. What is the original source of energy in our environment?

Questions 33 and 34 and refer to the following passage.

A student studies the animals in an environment and reports the following: There are trees and grasses. The rabbits and deer eat the grass, and the deer, squirrels, and small birds eat the fruits, berries, and leaves of the trees. Foxes hunt for the rabbits and squirrels. Large birds also eat the rabbits.

33. Copy the diagram below into your notebook. Fill in each of the organisms from the food web described above.

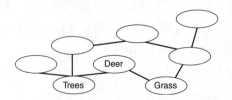

34. What might happen to this food web if a disease killed most of the rabbits?

35. Copy the table below into your notebook. Based on the food web on page 40, identify each of the organisms as a producer, herbivore, or carnivore by checking the appropriate box. The first organism has been done for you.

Organism	Producer	Herbivore	Carnivore
Grasses	✓		
Foxes			
Rabbits			
Deer			
Small birds			
Trees			
Large birds			
Squirrels			

36. Use the chart you completed in question 35 to draw a food pyramid for this food web.

Chapter 3

Living Systems: Biological Diversity and Heredity

Points to Remember

▷ Biologists classify organisms based on shared characteristics.

▷ Genetic information is passed from one generation to the next through chromosomes during reproduction.

▷ Mitosis produces two new cells with the same number of chromosomes as the original.

▷ Meiosis produces sex cells that contain half as many chromosomes as regular body cells.

▷ A Punnett Square can be used to predict the probability of inheriting a trait.

▷ A change in a gene is called a mutation. Mutations may contribute to a gradual change in a species. This process, called evolution, is usually very slow.

Classification

Placing Organisms in Groups

Since there are so many organisms, scientists must classify them in order to keep track of all the different types. A classification system groups things together by properties, which you choose. This is an important technique used by scientists in all areas of science.

Figure 3-1. *All things are classified as either living or nonliving. Identify the living and nonliving things in this illustration.*

If you were asked to classify the items in Figure 3-1 above into two groups, you might separate them as living things and nonliving things. If you look closely at the group of living things, you might separate it further into two smaller groups, plants and animals.

When you look at Figure 3-2, you might separate these animals into three types. This classification at first may seem obvious, but sometimes it can be very difficult. We are not always sure how to classify something.

Figure 3-2. *Scientists classify animals into different groups, such as mammals, birds, and fish. Where would the dolphin be placed?*

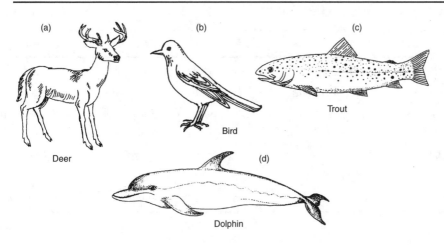

(a) Deer

(b) Bird

(c) Trout

(d) Dolphin

The three groups you might have chosen are mammals, birds, and fish. Where did you put the dolphin? The dolphin is a mammal. Scientists must carefully define the groups they use so that there is agreement on how to classify something. If the three groups you chose were animals that walk, animals that swim, and animals that fly, how would these groups be different? In which group would you place a penguin?

In 1737, a Swedish scientist named Carolus Linnaeus devised a classification system for living things. He grouped them based on internal and external structures and other shared characteristics. Linnaeus called the largest group a kingdom. Scientists now recognize five kingdoms of living things. The two kingdoms you are probably most familiar with are plants and animals. Each kingdom is broken down into subgroups called *phyla* (singular, *phylum*). Each phylum is broken down into classes, each class is broken down into orders, and each order is broken down into families. Each family is broken down into *genera* (singular, *genus*), and each genus is broken down into species. This may seem confusing at first, and you might wonder why we need so many different groups.

Think about how you address a letter. You include the country, state, city, zip code, street, and number. The country contains the largest number of locations. As you progress from state, to city, and eventually to street number, the areas get smaller and smaller. The system used by biologists works the same way. A kingdom contains an enormous number of different living things. As you progress down to species, the number of different kinds of living things in each group gets smaller and smaller.

Kingdom, phylum, class, order, family, genus, and species are assigned to every living thing (see examples in Table 3-1). We usually use only the last two names, the genus and species, to identify a living thing. The first letter of the genus is capitalized; the first letter of the species is not. Both are generally written in italics or are underlined. Together, the genus and species make up the scientific name.

The scientific name for a lion is *Panthera leo,* and for a house cat it is *Felis catus.* A lion and a house cat are both in the family Felidae, but they are different species. We say that they are both felines.

Table 3-1. *Classification of Living Things*

Group	House Cat	Red Maple	Lion	Human	Sugar Maple
Kingdom	Animalia	Plantae	Animalia	Animalia	Plantae
Phylum	Chordata	Tracheophyta	Chordata	Chordata	Tracheophyta
Class	Mammalia	Angiosperm	Mammalia	Mammalia	Angiosperm
Order	Carnivora	Dicotyledonae	Carnivora	Primates	Dicotyledonae
Family	Felidae	Aceraceae	Felidae	Hominidae	Aceraceae
Genus	*Felis*	*Acer*	*Panthera*	*Homo*	*Acer*
Species	*catus*	*rubrum*	*leo*	*sapiens*	*saccharum*

Multiple Choice

The table below gives the scientific classification for a variety of animals. Look at the table below and Table 3-1 to answer questions 1–4.

Group	Wolf	Dog	Horse	Grasshopper	Chimpanzee
Kingdom	Animalia	Animalia	Animalia	Animalia	Animalia
Phylum	Chordate	Chordate	Chordate	Arthropods	Chordate
Class	Mammalia	Mammalia	Mammalia	Insectae	Mammalia
Order	Carnivore	Carnivore	Ungulate	Orthoptera	Primates
Family	Canidae	Canidae	Equidae	Locustidae	Pongidae
Genus	Canis	Canis	Equus	Schistocerca	Pan
Species	lupus	familiaris	caballus	americana	troglodytes

1. Which animals are classified in the same genus but in different species?
    (1) dog and chimpanzee
    (2) horse and grasshopper
    (3) wolf and dog
    (4) wolf and horse

2. What is true about all the organisms in the table?
    (1) They all belong to a different kingdom.
    (2) They all belong to a different phylum.
    (3) They all belong to the same kingdom.
    (4) They all belong to the same phylum.

3. Which generally contains the largest number of different species?
    (1) phylum   (3) order
    (2) class   (4) genus

4. A particular species of grasshopper is most closely related to another organism that is in
    (1) the same phylum, but a different class
    (2) the same class, but a different order
    (3) the same order, but a different family
    (4) the same family, but a different genus

Thinking and Analyzing

5. Based on Tables in pages 44 and 45, which animal is most closely related to humans? Why?

6. Before an exam on classification, a student was repeating to himself, "King Philip Crossed Over From Germany to Spain." How would this sentence help him on his classification exam?

Reproduction

Cell Division

All cells come from other cells through the process of *cell division*. In this process, one "parent" cell divides into two new "daughter" cells. Every parent cell passes along to its daughter cells a set of "operating instructions" necessary for the cells to function properly. This genetic information is contained in threadlike structures called chromosomes, found in the cell nucleus.

The genetic information in the chromosomes also gives the cell, or the organism it belongs to, its individual characteristics, or traits, like size and shape. All members of a given species have the same number of chromosomes in each body cell. Chromosomes and their genetic information are passed on to the next generation during reproduction.

One-celled organisms reproduce through a kind of cell division called mitosis. In this process, a cell divides into two identical daughter cells, each of which contains the same number of chromosomes as the original parent cell (see Figure 3-3). When a one-celled organism undergoes mitosis, each new cell produced is a complete new organism.

In an organism made up of many cells, cells must duplicate themselves to build new tissue for growth and to repair damaged tissue. They do this

Figure 3-3. Cell division: mitosis produces two new cells with the same number of chromosomes as the original cell.

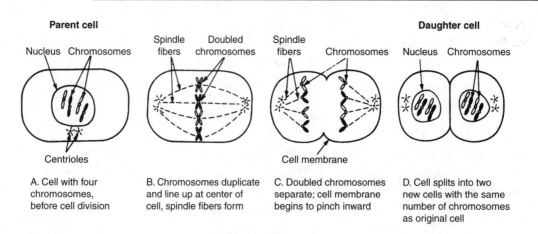

A. Cell with four chromosomes, before cell division

B. Chromosomes duplicate and line up at center of cell, spindle fibers form

C. Doubled chromosomes separate; cell membrane begins to pinch inward

D. Cell splits into two new cells with the same number of chromosomes as original cell

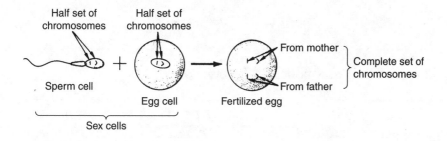

Figure 3-4. *Fertilization occurs when an egg and a sperm unite.*

through mitosis. Some of these organisms also produce offspring by mitotic cell division. In fact, any organism that reproduces asexually (with just one parent) does so through mitosis. The offspring created are genetically identical to the parent.

Sexual Reproduction

Some organisms reproduce sexually, with two parents. Sexual reproduction involves the joining of two special reproductive cells, one from each parent. These *sex cells,* or *gametes,* (sperm cells and egg cells) have only half the number of chromosomes held by other body cells. Sex cells are formed by a type of cell division called *meiosis.*

During sexual reproduction, a sperm cell from the male parent joins with an egg cell from the female parent. This is called **fertilization**. Since each sex cell contains half of a normal set of chromosomes, when they join, they form one cell with a complete set of chromosomes (Figure 3-4).

Process Skill

Interpreting Information in a Table

Many animals reproduce by the process of sexual reproduction. During sexual reproduction, sperm from the male joins with an egg from the female. This process, called fertilization, may take place either within the female's body (*internal fertilization*) or outside of the female's body (*external fertilization*). After the egg is fertilized, it develops into an embryo. This process may also take place either inside the female (*internal development*) or outside the female (*external development*).

For example, a chicken lays an egg, which is already fertilized. The chicken then sits on the egg to keep it warm as the embryo in the egg develops. Chickens have internal fertilization and external development. The table on page 48 shows some different animals, their classes and habitats, and the types of fertilization and development they undergo.

(Continued)

Study the table and answer questions 1–5 below.

Animal	Class	Habitat	Type of Fertilization	Type of Development
Goldfish	Osteichthyes (Bony fishes)	Water	External	External
Bluebird	Aves (Birds)	Land	Internal	External
Bee	Insecta	Land	Internal	External
Dog	Mammalia	Land	Internal	Internal
Frog	Amphibia	Water and Land	External	External
Lizard	Reptilia	Land	Internal	External
Whale	Mammalia	Water	Internal	Internal

QUESTIONS

1. Based on the table, what is required for external fertilization?
   (1) internal development
   (2) land habitat
   (3) water habitat
   (4) only fish can have external fertilization
2. Which of the following is required for internal development?
   (1) land habitat   (3) external fertilization
   (2) water habitat   (4) internal fertilization
3. A salamander belongs to the class Amphibia. What would you predict about salamanders?
   (1) They live both on land and in water.
   (2) They have internal development.
   (3) They live in water only.
   (4) They have internal fertilization.
4. Alligators have internal fertilization and external development. Based on the table, which class do they most likely belong to?
   (1) Amphibia   (2) Reptilia   (3) Fish   (4) Mammalia
5. Based on this chart, what two generalizations can be made about mammals?

Development of the Fertilized Egg

The fertilized egg cell, called the *zygote*, forms new identical cells through mitosis. We know however, that multicellular organisms contain may different types of cells. (See Figure 4-1 on page 62.) At a certain stage in the development of the organism, the cells arrange themselves into different layers. Within these layers, the cells begin to change into the tissues and organs of the body. This process, called *differentiation*, is illustrated in Fig-

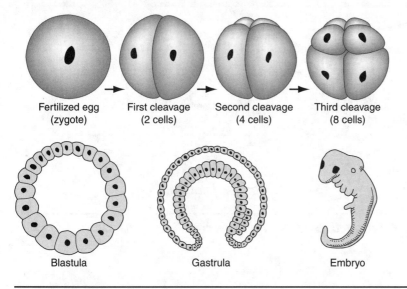

Figure 3-5. *Cleavage from zygote to embryo.*

ure 3-5, which shows the development from the one-celled zygote to the multicellular embryo. Eventually, an organism develops so that it resembles its parents. However, the new organism is not identical to either parent, but has traits from both. In this way, sexual reproduction leads to variation in the next generation. Figure 3-6 shows a possible result of sexual reproduction in chickens.

Sexual Reproduction in Plants

Plants, like animals, may reproduce sexually. However, since plants are not mobile, they have developed different structures and techniques to accomplish this life process.

Figure 3-6. *Sexual reproduction leads to variation in offspring.*

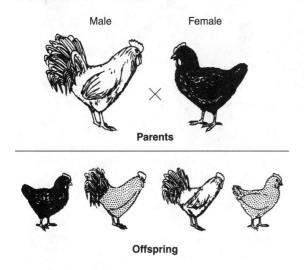

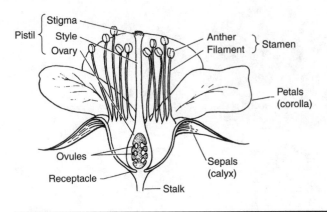

Figure 3-7. *Reproductive parts of a flower.*

A flower is the reproductive organ of a plant. Pollen produced in the *anther* (see Figure 3-7) contains the male sperm cells. The female egg cells are located in the ovary at the base of the *pistil*. In plants, as in animals, the male and female reproductive cells must join in order for fertilization to take place. A plant depends on wind, rain, insects, birds or other small animals to transfer the pollen from the anther to the pistil. Some plants are completely dependent on bees as their means of fertilization.

The fertilized egg develops into an embryo. The seed is a container that houses the embryo and provides food for its early development. The ovary of the flower develops into a fruit. Animals that eat the fruit help distribute the seed to new locations. This is called *seed dispersal*. Wind is another agent of seed dispersal. Different types of seeds have developed to take advantage of various methods of dispersal, as illustrated in Figure 3-8.

The embryo, contained in the seed, begins to sprout into a young plant when conditions are favorable. Generally, the most important conditions are

Figure 3-8. *Methods of seed dispersal for different types of seeds.*

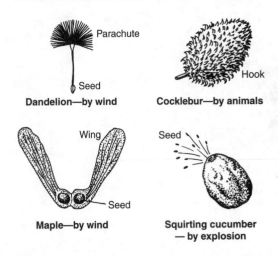

temperature and moisture. The sprouting process is called **germination**. During germination, the plant's roots, stem and leaves begin to develop.

Multiple Choice

7. Variation in a new generation of organisms is the result of

    (1) sexual reproduction involving one parent

    (2) sexual reproduction involving two parents

    (3) asexual reproduction involving one parent

    (4) asexual reproduction involving two parents

8. Which of the following diagrams shows sexual reproduction?

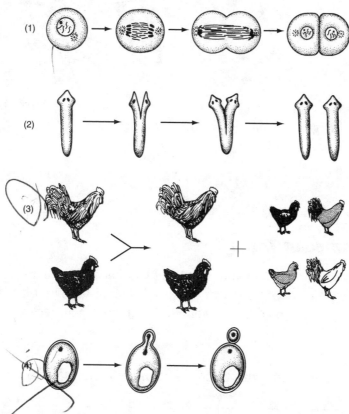

9. Egg and sperm cells are formed by a process called

    (1) mitosis   (2) fertilization   (3) asexual reproduction   (4) meiosis

10. What process is illustrated by the following :

    egg + sperm → zygote

    (1) meiosis   (2) mitosis   (3) fertilization   (4) cleavage

Thinking and Analyzing

11. A human cell contains 46 chromosomes. How many chromosomes will each daughter cell contain after mitosis has taken place?

12. A fruit fly has 4 chromosomes in each of its sperm cells. How many chromosomes does it have in its regular body cells?

Base your answer to questions 13 and 14 on the diagram below which shows a form of reproduction.

13. Which type of reproduction is shown in the diagram?

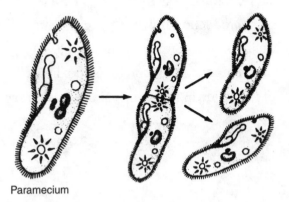

Paramecium

14. How does the genetic material of each of the daughter cells compare with the genetic material of the parent cell?

Heredity

Inheritance of Traits

What type of earlobes do you have? Did you know that there are two types of earlobes, "attached" and "free"? (See Figure 3-9.) The type of earlobes you have is determined by the genetic information you received from each

Figure 3-9. The two types of earlobes that can be inherited: attached and free.

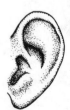

Free earlobe   Attached earlobe

parent. A piece of genetic information that influences a trait is called a gene. You received one gene containing information for the type of earlobes you have from each of your parents. These two genes can be the same or different.

If both genes match for a trait, that same trait will appear, and you are said to be "pure" for that trait. However, if you have differing genes for a trait, one that says "attached" and another that says "free," you are said to be "hybrid" for that trait.

The type of earlobe is just one of thousands of traits you inherit from your parents. Each of these traits is controlled by one or more pair of genes. Therefore, each of your body cells contains thousands of different genes.

Mendelian Genetics

An Austrian monk named Gregor Mendel (1822–1884) performed experiments with pea plants to investigate how traits are inherited. He determined that for traits controlled by two differing genes (which he referred to as "factors"), one of these genes would be dominant over the other. An individual with both genes (a hybrid) will exhibit only the characteristics of the dominant gene. The gene for the trait that is not exhibited is called a recessive gene.

Mendel crossed pure tall pea plants with pure short pea plants. He observed that all the offspring were tall. He concluded that tall is the dominant trait. He also concluded that all of the offspring of this generation must be hybrids. When the hybrid tall pea plants were crossed with one another, most of the offspring were tall, but some were short. This proved that the hybrid tall pea plants still contained a gene (factor) for shortness. Only when the offspring received two short genes did they appear as short plants. Table 3-2 indicates some common dominant and recessive traits studied in genetics.

Table 3-2. *Common Traits Studied in Genetics*

Organism	Trait	Dominant	Recessive
Human	Eye color	Brown	Blue
Human	Earlobe	Free	Attached
Human	Blood type	A or B	O
Fruit fly	Wing	Normal	Vestigial
Fruit fly	Eye color	Red	White
Pea plant	Height	Tall	Short
Pea plant	Pea color	Yellow	Green
Pea plant	Seed shape	Round	Wrinkled

Table 3-3. *Possible Gene Combinations*

Type of Genes	Representation	Appearance
Pure red	RR	Red
Hybrid red	Rr	Red
Pure white	rr	White

Using a Punnett Square

Fruit flies are often used to study genetics because they mature and reproduce quickly. Several generations can be observed in a short period of time.

One fruit fly characteristic studied is eye color. Fruit flies can have red eyes or white eyes. Red is dominant while white is recessive. We can predict the possibility of an offspring having red eyes or white eyes if we know the types of genes in the parents. To represent the gene, we use a capital letter for the dominant trait (R for red) and a lowercase of the same letter for the recessive trait (r for white). The possible gene combinations are outlined in Table 3-3.

A diagram called a **Punnett square** can be used to predict the probability of an organism inheriting a given trait. The genes for one parent are placed at the top of the square and the genes for the other parent are placed at the side. Below is a Punnett square for two hybrid parents.

	R	r
R	RR	Rr
r	Rr	rr

Key:

R = gene for red eyes

r = gene for white eyes

Each box is filled in with the letter appearing above it and to its left. These boxes represent the possible combinations of genes. We see that of the four boxes, one will be pure dominant (RR), one will be pure recessive (rr), and two will be hybrid (Rr). This can be summarized as shown in Table 3-4.

Table 3-4. *Probabilities of Offspring of Hybrid Red-Eyed Fruit Flies*

Type of Genes	Appearance	Probability
RR-pure dominant	Red eyes	1/4 (25%)
Rr-hybrid dominant	Red Eyes	2/4 (50%)
rr-pure recessive	White eyes	1/4 (25%)

Process Skill

Interpreting the Results of an Experiment

Indira studied the traits of an unusual insect. She noticed that some of the insects had long antennae and some had short antennae. She separated the insects into two groups based on their antennae sizes and allowed them to mate only within each group. After the eggs were laid, the parents were removed. The eggs were allowed to develop. In one group, all the offspring had short antennae. Offspring in the other group had both types of antennae.

QUESTIONS

1. Indira can tell from this information that
   (1) short is dominant
   (2) long is recessive
   (3) short is recessive
   (4) there were two different species of insects
2. In describing the parents, it would be safe to infer that
   (1) all those with long antennae were pure
   (2) all those with long antennae were hybrid
   (3) some of those with long antennae were hybrid
   (4) some of those with short antennae were hybrid
3. To improve this experiment, which of the following should not be done?
   (1) Separate the insects to be mated as soon as they are hatched.
   (2) Feed the long- and short-antennae groups different diets.
   (3) Maintain a constant environment for both groups.
   (4) Repeat the experiment several times.
4. Predict the results of a cross between a pure long-antennae insect and a pure short antennae insect.
   (1) Some of the offspring will have short antennae.
   (2) All of the offspring will have short antennae.
   (3) Some of the offspring will have long antennae.
   (4) All of the offspring will have long antennae.

Review Questions

Multiple Choice

15. Based on Table 3-4, what percentage of the fruit fly offspring are likely to have red eyes?
    (1) 25%   (2) 50%   (3) 75%   (4) 100%
16. What fraction of the red-eyed flies produced are pure for the trait?
    (1) 1/2   (2) 1/4   (3) 1/3   (4) 2/3

17. Brown eyes are dominant over blue eyes in humans. What is the likelihood of two blue-eyed parents having a brown-eyed child?

(1) 0%   (2) 25%   (3) 50%   (4) 100%

Thinking and Analyzing

18. Some diseases cannot be caught because they are inherited. Disorders that fall into this category include Tay-Sachs disease, sickle-cell anemia, hemophilia, and cystic fibrosis. A child can suffer from one of these diseases only if both parents carry the gene for the disease. What can you conclude about the genetics of these diseases? only
~~worth an kid when two corect both?~~

 parents.

19. A farmer wanted to get rid of a recessive gene from a population of livestock. For generation after generation, she bred only animals that showed the dominant trait. Would she be successful? Explain.

Process Skill

Interpreting a diagram

A diagram that shows how a trait is passed from generation to generation is called a *pedigree*. For questions 1–5, use the pedigree below to determine the genetic makeup of the individuals represented.

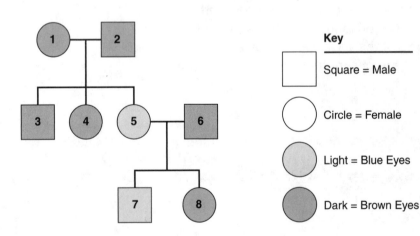

Key

Square = Male

Circle = Female

Light = Blue Eyes

Dark = Brown Eyes

QUESTIONS

1. Individual 4 can best be described as
   (1) a brown-eyed male   (3) a blue-eyed male
   (2) a brown-eyed female   (4) a blue-eyed female

(Continued)

2. Based on this pedigree, it would be safe to assume that
  (1) blue eyes are dominant   (3) brown eyes are recessive
  (2) blue eyes are recessive   (4) eye color is not inherited
3. The genetic makeup of individual 2 can best be described as
  (1) BB   (2) Bb   (3) bb
4. How do we know that individual 6 must be hybrid for eye color?
5. Use a Punnett Square to predict the probability of the next child of individuals 5 and 6 having blue eyes.

Evolution

Mutations

Evolution accounts for the great diversity among living things. The evolutionary process depends on the passing of genetic information from one generation to the next through chromosomes. This occurs during reproduction.

If parents and offspring are so similar, why are there so many adaptations and such diversity among living things? Genetic accidents do occur. Sometimes genetic material does not reproduce properly. This may be caused by a natural "accident" or by something in the environment. Such a genetic accident is called a **mutation**. The new genetic information will cause a variation in the offspring. If this change is harmful to the organism, it will be less likely to survive and reproduce. If the change is beneficial to the organism, it will be better able to survive and reproduce. The new genetic information will then be passed on to each new generation. If these changes increase the likelihood that the organism will reproduce, the changes will become more common within the population.

Scientists have discovered that mutations can be caused by environmental factors. Forms of radiation such as gamma rays, X rays and even ultraviolet light can cause mutations. In addition, certain chemicals have been shown to cause changes in genetic material.

Natural selection favors those organisms that are best able to survive and reproduce. After many generations, and many mutations, the organism may look and behave so differently from its ancestors that it has become a new species. This process, called *evolution*, is often very slow and may take millions of years. However, bacteria and insects with very short life cycles may evolve relatively quickly. For example, penicillin is no longer effective against certain bacteria. Differences in genetic makeup may allow an individual, resistant bacterium, to survive treatment with penicillin. It will then, reproduce, to form additional, resistant bacteria. These penicillin-resistant bacteria may soon become the main variety causing the infection. The infection can no longer be treated with penicillin. New antibiotics must be developed to kill the new resistant varieties. Similarly, certain insect populations have become resistant to certain insecticides.

Today, through advances in genetic engineering, scientists can cause a species to change to improve a particular characteristic. Plants and livestock with desirable traits such as resistance to disease, better taste, and higher yields are chosen in a process called *selective breeding*. Selective breeding is a process in which individuals with the most desirable traits are crossed, or allowed to mate, with the hopes that their offspring will show the desired traits. For example, turkeys have been selectively bred to produce more white meat, racehorses are selectively bred for speed, and wheat is selectively bred to produce larger heads of grain. Through selective breeding, humans have replaced nature in determining which organisms survive and reproduce.

Genetic Diseases

Many mutations are not beneficial. Diseases, such as sickle-cell anemia or hemophilia, and conditions such as albinism (a lack of pigment) are caused by genetic changes that are passed down from generation to generation. Scientists have developed ways of checking genetic material for the presence of some of these harmful genes. Since most of these diseases are caused by recessive genes, both parents must carry the gene for the child to be at risk. A new health care field called genetic counseling has arisen to help parents understand and evaluate the risk that their children may inherit one of these diseases.

DNA

Chromosomes contain complex molecules called DNA. DNA consists of smaller molecules that are arranged in a particular sequence. This sequence of molecules provides a code that determines the genetic information.

In a coordinated national effort, called the Human Genome Project, scientists are decoding the messages contained in the chromosomes. They hope to use this information to predict and even cure genetic diseases.

Genetic Engineering

Scientists have learned how to deliberately alter the genetic material of an organism to benefit humankind. They have developed new strains of plants and animals that are resistant to disease and yield more food. This is done by actually changing the genetic material in these organisms in a process called *genetic engineering*. Once these changes have been made, they are passed on to all future generations. For example, sheep have been genetically engineered to produce human insulin, a hormone that is used to treat diabetes. However, some people feel that genetic engineering should not be attempted because the resulting organisms might turn out to be more harmful than beneficial. In addition, tremendous controversy has arisen over the morality of genetically altering human beings.

 ## Review Questions

Multiple Choice

20. After exposure to X rays, fruit flies sometimes produce offspring with unusual traits such as white eyes or short, useless wings. This can best be explained by

 (1) the effect of X rays on the eyes and wings of fruit flies

 (2) the effect of X rays on the genes of fruit flies

 (3) the effect of X rays on the ability of fruit flies to reproduce

 (4) the relationship between eye color and wing size

21. The process described in question 20 is called

 (1) fertilization   (3) a mutation

 (2) a life cycle   (4) an adaptation

22. The process of deliberately changing a trait in an organism in order to benefit humans is known as

 (1) adaptation   (3) natural selection

 (2) mutation   (4) genetic engineering

Thinking and Analyzing

Read the passage below and use the information to help you answer questions 23–25.

In the early 19th century, Jean Baptiste Lamarck, attempted to explain how organisms change over time. He believed that giraffes evolved from short-necked ancestors. These ancestors needed to stretch their necks to reach leaves that were high up in the trees. According to Lamarck, a lifetime of stretching to reach food caused their necks to become longer. The offspring inherited the slightly longer necks of the parents. Over many generations, the giraffe's neck reached its current length. Lamarck's theory was called "The Law of Inheritance of Acquired Characteristics."

In an experiment to verify The Law of Inheritance of Acquired Characteristics, scientists cut off the tails of mice, and allowed them to reproduce. They repeated this process on all of the offspring, for more than 20 generations. All of the mice were born with tails just like those of the original parents.

23. The experiment on the mice

    (1) supported Lamarck's Theory, because the acquired trait was inherited.

    (2) disproved Lamarck's Theory, because the acquired trait was inherited.

    (3) supported Lamarck's Theory, because the acquired trait was not inherited.

    (4) disproved Lamarck's Theory, because the acquired trait was not inherited

24. If the experiment on the mice were carried out for several more generations, what would you predict about the lengths of their tails?

    (1) They would have no tails.

    (2) They would have tails as long as their parents'.

    (3) They would have short tails.

    (4) They would have white tails.

25. The modern theory of evolution would explain the change in the giraffe's appearance by using all of the following terms except

    (1) mutation   (3) natural selection

    (2) variation   (4) acquired characteristics.

Read the passage below and use it to help you answer questions 27–28.

There is a popular joke, "Insanity is inherited; you get it from your children." In reality, many diseases are inherited; you get them from your parents. Before having children, adults with genetic diseases may seek the advice of a genetic counselor.

26. Genetic counselors may provide information on all of the following except

    (1) The probability of a disease being inherited

    (2) Vaccinations to prevent these diseases

    (3) Checking genetic material for the presence of harmful genes

    (4) The risks and dangers associated with the disease

27. A genetic counselor informs parents that the disease they are inquiring about is carried on a recessive gene and that only one of the parents carries this gene. What can the parents conclude about the health of their future children?

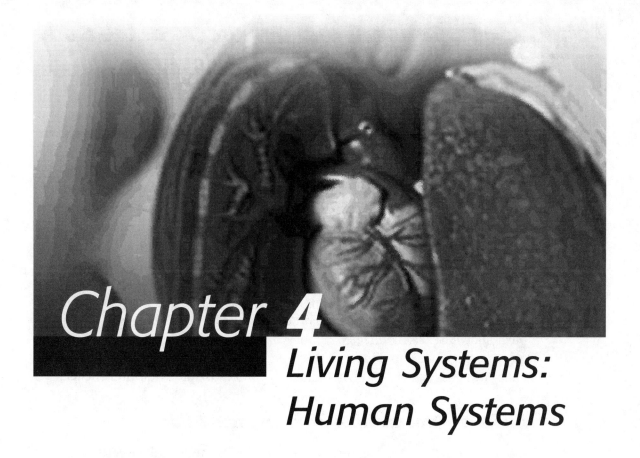

Chapter 4
Living Systems: Human Systems

Points to Remember

▷ Complex organisms, such as humans, show several levels of body organization.

▷ Cells are the basic unit of life. Tissues are groups of similar cells that work together to carry out a life process. Organs are groups of tissues that work together to carry out a life process.

▷ Organ systems are groups of organs that work together to carry out a life process. Different organ systems work together to carry out life processes.

▷ The human body includes the following types of tissues: blood, bone, muscle, nerve, and skin.

▷ The human body includes the following systems: skeletal, muscular, nervous, endocrine, digestive, circulatory, respiratory, excretory, and reproductive.

▷ All the body systems are interdependent; for example, the nervous system and endocrine system act together to regulate and control body activities.

▷ The circulatory system and respiratory system work together to bring oxygen in the blood to all body cells.

Organization, Support, and Movement of the Body

Human Body Systems Are Interdependent

A human being is a complex organism, made up of a number of different body systems. Each system carries out a specific life process, and thereby contributes to the operation of the body as a whole.

In addition, all body systems are interdependent, and work with one another to keep a person alive. For instance, the respiratory system brings needed oxygen into the body; the oxygen is then transported throughout the body by the circulatory system.

All of the systems of the body work together to maintain a balanced internal state (homeostasis). Diseases as well as personal behaviors, such as poor dietary habits and the use of toxic substances, (alcohol and tobacco) may interfere with the ability to maintain this balance. Some of the effects on the body may be immediate while others may not appear for years. During pregnancy, diseases contracted by the mother may also affect the development of the child. Smoking and drinking alcoholic beverages during pregnancy decrease the birth weight of the baby and cause other harmful effects.

Levels of Organization in the Human Body

1. *Cells.* Living things are made up of basic units called ***cells***. The human body contains many types of cells; each designed to perform a different function. Figure 4-1 shows several kinds of cells.

2. *Tissues.* A group of similar cells acting together forms a ***tissue***. Skin tissue covers the body. Muscle tissue produces body movements. Table 4-1 on page 63 lists some types of human tissues.

Figure 4-1. *Different types of cells.*

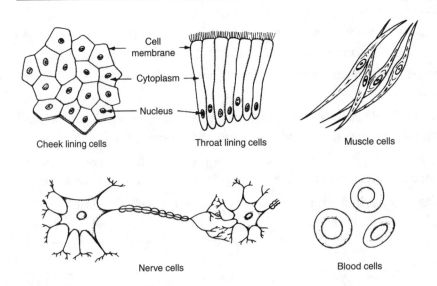

Cell membrane

Cytoplasm

Nucleus

Cheek lining cells

Throat lining cells

Muscle cells

Nerve cells

Blood cells

Table 4-1. *Types of Human Tissue and Their Functions*

Tissue	Function
Blood	Transports materials throughout the body
Bone	Supports and protects body and organs
Muscle	Helps body to move; aids in circulation, digestion, and respiration
Nerve	Carries messages
Skin	Covers and protects body; excretes wastes

Table 4-2. *Important Organs and Their Functions*

Organ	Function
Heart	Pumps blood
Kidney	Removes wastes from blood
Lung	Exchanges gases with the environment
Stomach	Breaks down food by physical and chemical means
Brain	Controls thinking and voluntary actions

3. *Organs.* A group of tissues working together forms an **organ**. The heart is an organ that pumps blood throughout the body. It is composed mainly of muscle tissue, but also contains blood tissue and nerve tissue. Table 4-2 above lists some important organs.

4. *Organ Systems.* A group of organs acting together to carry out a specific life process makes up an **organ system**. The circulatory system is a system that carries out the process of transport, moving materials throughout the body. Table 4-3 lists the human organ systems.

Table 4-3. *Human Organ Systems*

System	Function	Examples of Organs or Parts
Skeletal	Supports body, protects internal organs	Skull, ribs
Muscular	Moves organs and body parts	Arm and leg muscles
Nervous	Controls body activities; carries and interprets messages	Brain, spinal cord
Endocrine	Regulates body activities with hormones	Adrenal, pituitary glands
Digestive	Breaks down food into a usable form	Stomach, intestines
Circulatory	Carries needed materials to body cells and waste materials away from cells	Heart, arteries, veins
Respiratory	Exchanges gases with the environment	Lungs, bronchi
Excretory	Removes wastes from the body	Kidneys, skin
Reproductive	Produces offspring	Ovaries, testes

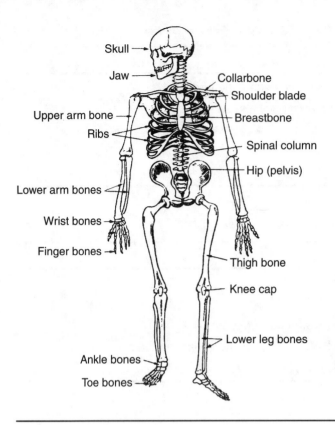

Figure 4-2. *The human skeleton.*

The Skeletal System

The human *skeletal system,* shown in Figure 4-2, supports and protects the body and its organs. The skeletal system includes the *skull, spinal column, breastbone, ribs,* the *bones of the limbs* (arms and legs), and *cartilage.*

1. *Bones and Cartilage.* **Bones** are made of hard, strong material. **Cartilage** is a softer, more flexible tissue. Cartilage acts as a cushion between bones, and provides flexibility at the ends of bones. Disks of cartilage separate the bones of the spinal column, cushioning them from one another.

2. *Joints.* Where one bone is connected to another bone, a **joint** is formed. Most joints, such as the knee and elbow, allow the bones to move. However, some joints, like those in the skull, do not allow movement. Figure 4-3 shows three types of joints.

3. *Ligaments and Tendons.* At movable joints, the bones are held together by strips of tissue called *ligaments.* Bones are moved by muscles, which are attached to bones by *tendons,* cordlike pieces of tissue. A common sports injury is a torn Achilles tendon, which joins the calf muscles of the leg to the heel bone.

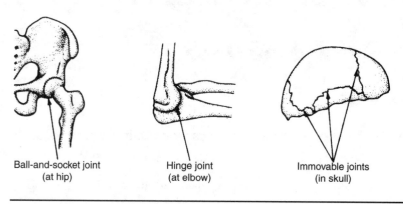

Ball-and-socket joint
(at hip)

Hinge joint
(at elbow)

Immovable joints
(in skull)

Figure 4-3. *Three types of joints.*

The Muscular System

Muscles are masses of tissue that contract to move bones or organs. The *muscular system* contains two main kinds of muscles: *voluntary* and *involuntary*.

1. *Voluntary Muscles.* The *skeletal muscles*, which move bones, are examples of **voluntary muscles**—muscles that are controlled by our will. Skeletal muscles work with the skeleton to move body parts (see Figure 4-4), and thereby produce locomotion. **Locomotion** is the movement of the body from place to place. The muscles in the face and around the eyes are also voluntary muscles.

2. *Involuntary Muscles.* **Involuntary muscles** are not under our conscious control. There are two types of involuntary muscles: cardiac and smooth. *Cardiac* muscle, present only in the heart, pumps blood throughout the body. *Smooth* muscle, found in the respiratory, circulatory, and digestive systems, aids in breathing, controlling blood flow, and movement of food.

Figure 4-4. *Muscles, tendons, and bones of the arm enable movement.*

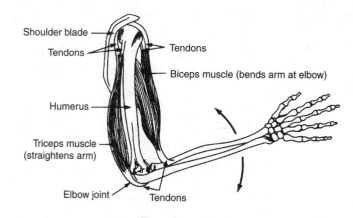

Shoulder blade

Tendons

Tendons

Biceps muscle (bends arm at elbow)

Humerus

Triceps muscle
(straightens arm)

Elbow joint

Tendons

Review Questions

Multiple Choice

1. A group of organs working together to carry out a life process is called
   (1) a cell
   (2) a tissue
   (3) an organ
   (4) an organ system

2. A group of cells acting together makes up
   (1) a cell
   (2) a tissue
   (3) an organ
   (4) an organ system

3. A tissue designed to carry messages throughout the body is most likely to be
   (1) skin   (2) muscle   (3) nerve   (4) bone

4. Going from the simple to the complex, which order correctly represents the organization of the human body?
   (1) organ system → organ → cell → tissue
   (2) cell → tissue → organ → organ system
   (3) tissue → cell → organ → organ system
   (4) organ → organ system→ cell → tissue

5. Which body system supports and protects other body systems?
   (1) skeletal
   (2) endocrine
   (3) reproductive
   (4) digestive

6. Which body system is mainly responsible for the movement of the body?
   (1) digestive
   (2) circulatory
   (3) muscular
   (4) endocrine

7. Which group lists three parts of the skeletal system?
   (1) heart, stomach, brain
   (2) tendons, nerves, brain
   (3) bones, nerves, blood
   (4) cartilage, ligaments, bones

8. Which type of muscle is found only in the heart?
   (1) voluntary
   (2) smooth
   (3) cardiac
   (4) involuntary

9. Which activity is most likely to be controlled by a smooth muscle?
   (1) breathing   (2) walking   (3) chewing   (4) thinking

10 The diagram below best demonstrates that

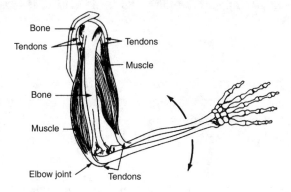

(1) the skeleton protects body organs

(2) bones are held together at joints by ligaments

(3) muscles and bones work together to move body parts

(4) cartilage protects and cushions bones

Thinking and Analyzing

11 An outer layer of hard material called chitin protects the body of an insect. This material gives the insect its shape, protects its organs and gives the insect support. Which system in humans carries out the same function?

12 The organs of a plant include the stems, leaves, roots and flowers as illustrated in the diagram below. Which structure in the plant is most similar in functions to the skeletal system in giraffes.

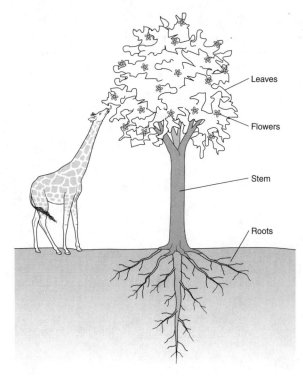

Regulation, Digestion, and Circulation

Regulation

The *nervous system* and the *endocrine system* work together to regulate body processes and actions. They provide us with a way of detecting and responding to *stimuli* (changes inside or outside the body). The nervous system (Figure 4-5) is made up of the *brain, spinal cord, nerves,* and parts of the *sense organs*.

1. The **brain** receives and interprets *nerve impulses* ("messages"), and controls thinking, voluntary action, and some involuntary actions, such as coordination, balance, breathing and digestion.
2. The **spinal cord** channels nerve impulses to and from the brain, and controls many automatic responses, or *reflexes*, such as pulling your hand away from a flame.
3. **Nerves** provide a means of communication between the sense organs, the brain and spinal cord, and muscles and glands.
4. The **sense organs**, which include the skin, eyes, ears, nose, and tongue, receive information from the environment.

Nerve cells, also called **neurons**, receive and transmit nerve impulses (see Figure 4-6). Two types of neurons are the sensory and the motor neurons. *Sensory neurons* carry information from the sense organs to the brain or spinal cord. *Motor neurons* carry messages from the brain or spinal cord to muscles and glands, which respond to the messages.

Figure 4-5. The human nervous system.

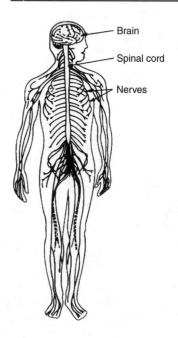

Brain

Spinal cord

Nerves

Figure 4-6. A typical neuron, or nerve cell.

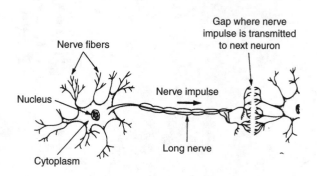

Gap where nerve impulse is transmitted to next neuron

Nerve fibers

Nerve impulse

Nucleus

Cytoplasm

Long nerve

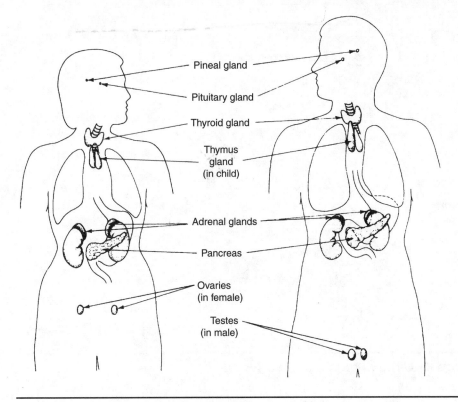

Figure 4-7. *Some major glands of the human endocrine system.*

The endocrine system is made up of glands. A **gland** is an organ that makes and *secretes* (releases) chemicals. Endocrine glands secrete chemicals called **hormones**. Figure 4-7 shows some of the endocrine glands. When an endocrine gland secretes a hormone into the bloodstream, the blood carries the hormone to an organ, which responds in some way. For example, if you are suddenly faced with some danger, such as a snarling dog, the hormone *adrenaline* is released by your *adrenal gland*. The adrenaline makes your heart beat faster and your breathing more rapid. More sugar is released into your bloodstream to provide energy. These changes prepare your body to respond to the danger.

The Digestive System

Our cells need nutrients from food for energy, growth, and repair. The *digestive system* breaks down food into soluble nutrients that can then be absorbed into the bloodstream and carried to the cells.

The digestive system, shown in Figure 4-8 on page 70, consists of the digestive tract and the accessory organs.

1. The *digestive tract* is a tube in which food travels through the body. It begins at the mouth and continues through the *esophagus, stomach, small intestine,* and *large intestine.*
2. The *accessory organs* are the *pancreas, gallbladder,* and *liver.* They produce digestive juices that are released into the digestive tract.

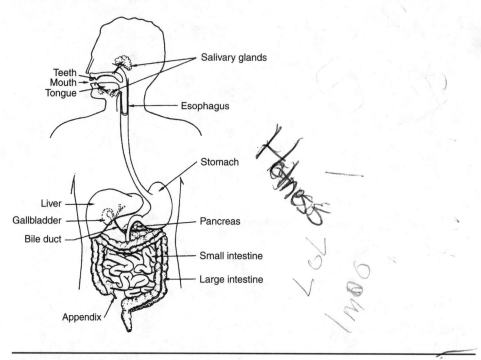

Figure 4-8. *The human digestive system.*

Table 4-4 lists the digestive juices, where they are produced, and what foods they digest.

The digestive system breaks down food by *physical* and *chemical* means. (1) Food is physically broken down into small bits by chewing and by the action of muscles in the digestive tract. (2) The chemical breakdown of food releases nutrients that can be used by cells. Chemicals called enzymes, found in the digestive juices accomplish this breakdown.

Digestion starts in the mouth, and continues in the stomach and small intestine. When digestion has been completed, digested materials are absorbed into the bloodstream through the walls of the small intestine. Undigested materials, which make up the solid wastes called *feces*, pass on through the large intestine and are expelled from the body.

Table 4-4. *Digestive Juices*

Organ	Digestive Juice	Foods Acted On
Mouth	Saliva	Starches
Stomach	Gastric juice	Proteins
Small Intestine	Intestinal juices	Sugars, proteins, fats
Pancreas	Pancreatic juice	Proteins, starches, fats
Liver	Bile	Fats

Note: Bile and pancreatic juice are secreted by the liver and pancreas *into* the small intestine, where digestion occurs.

The Circulatory System

Nutrients absorbed into the blood must be transported to all body cells. This is the job of the *circulatory system*: to bring needed materials such as nutrients, water, and oxygen to the cells and to carry away wastes, like carbon dioxide, from the cells.

The components of the circulatory system are the blood, the heart, the blood vessels (arteries, veins, and capillaries), lymph, and lymph vessels.

1. *Blood.* The **blood** is a liquid tissue containing plasma, red and white blood cells, and platelets. Plasma is the liquid part of the blood that carries dissolved nutrients, wastes, and hormones. The red blood cells, the white blood cells and the platelets float in the plasma. Red blood cells contain the pigment *hemoglobin*, which is the chemical that carries oxygen to the cells. White blood cells fight infection. They surround the infecting organism and destroy it. Platelets cause clotting, which stops bleeding.

2. *Heart.* The **heart** (Figure 4-9) is a muscle that contracts regularly to pump blood throughout the body. The blood is pumped from the heart to the lungs, where it receives oxygen and gets rid of carbon dioxide. The blood then returns to the heart to be pumped to the rest of the body, as shown in Figure 4-10 on page 72.

3. *Blood Vessels.* The blood flows through a network of tubes called **blood vessels**. There are three types of blood vessels. **Arteries** carry blood away from the heart, while **veins** return blood to the heart. Connecting arteries to veins are *capillaries*. Through the walls of the extremely small capillaries, essential materials are exchanged between the blood and the body's cells. Dissolved nutrients, water, and oxygen pass from the blood into the cells, and some wastes from the cells pass into the blood.

4. *Lymph.* Some of the watery part of the blood filters out through the walls of the capillaries into the surrounding tissue. This fluid, called

Figure 4-9. *The human heart.*

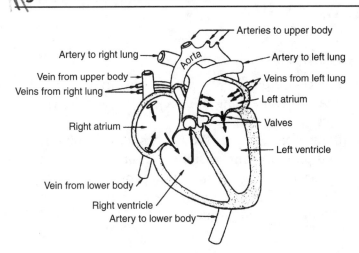

Arteries to upper body

Artery to right lung — Aorta — Artery to left lung

Vein from upper body — Veins from left lung

Veins from right lung —

Left atrium

Right atrium — Valves

Left ventricle

Vein from lower body —

Right ventricle —

Artery to lower body —

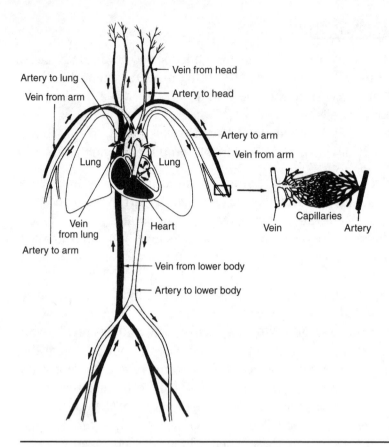

Figure 4-10. Pathways of blood through the human circulatory system.

lymph, bathes all the cells of the body. Lymph acts as a go-between in the exchange of materials between the blood and the cells. After receiving wastes from the cells, lymph is collected and returned to the bloodstream through *lymph vessels*.

 *Laboratory Skill*

Doing an Experiment

In the movies, how does a police officer determine if a person lying in the street is still alive? You may have noticed the officer touching the neck or wrist of the victim. The officer is checking for a pulse.

A pulse is an observable throbbing produced as the heart pushes the blood through an artery. You may find your pulse in your wrist, neck, thumb, or temple. You can determine how fast your heart is beating by measuring your pulse rate, the number of beats per minute. How does pulse rate change as the result of exercise? Jerry and Pauline did an experiment to find out.

(Continued)

Part I—Resting pulse rate.

Before taking any measurements, they relaxed in a sitting position for two minutes. Using their fingertips (as seen in the figure), they found their pulse in their wrists. They each used a stopwatch to time 15 seconds while they counted the number of beats during that time period. (They multiplied the number of beats by four to determine the number of beats per minute.)

Part II—Pulse rate after exercise.

Pauline and Jerry jogged in place for two minutes. They sat down and immediately took their pulses for 15 seconds. The results are record below.

Results: *Copy the table below into your notebook for your results:*

	Number of Beats in 15 Seconds	
	Pauline	Jerry
Before exercise	18	21
After exercise	27	33

QUESTIONS

1. Calculate Pauline's resting pulse in beats per minute.
2. What effect does exercising have on the pulse rate?
3. Explain why exercising might change the pulse rate.
4. State a hypothesis that might be tested by this experiment.

Review Questions

Multiple Choice

13. Hormones are chemicals secreted by the
    (1) gall bladder
    (3) endocrine glands
    (2) brain
    (4) small intestine

14. The endocrine system works with the nervous system to
    (1) digest nutrients
    (2) exchange gases with the environment
    (3) produce energy
    (4) regulate body activities

15. The brain, spinal cord, and sensory neurons are all part of the

(1) nervous system   (3) circulatory system

(2) respiratory system   (4) endocrine system

16. The human cell shown is designed to

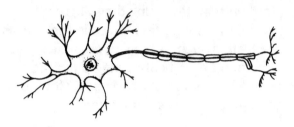

(1) store excess food

(2) send and receive nerve impulses

(3) cover and protect the body

(4) carry oxygen to other cells

17. Food is broken down into a usable form by the

(1) nervous system   (3) digestive system

(2) skeletal system   (4) circulatory system

18. Which group includes structures that are all parts of the digestive system?

(1) heart, lungs, pituitary gland

(2) adrenal gland, pituitary gland, thyroid gland

(3) skin, kidneys, lungs

(4) stomach, intestines, pancreas

19. Solid materials that are not digestible are eliminated from the body as

(1) urine   (2) perspiration   (3) lymph   (4) feces

20. The function of the circulatory system is to

(1) carry materials to and from the cells

(2) break down food into a usable form

(3) regulate body activities

(4) respond to stimuli

21. Which group includes structures that all belong to the circulatory system?

(1) heart, liver, and lungs

(2) arteries, veins, and capillaries

(3) arteries, kidneys, and stomach

(4) bones, cartilage, and ligaments

22. Which represents the correct pathway of the nutrients in an apple once you take a bite?

(1) circulatory system → cell → digestive system

(2) cell → digestive system → circulatory system

(3) digestive system → circulatory system → cell

(4) circulatory system → digestive system → cell

Thinking and Analyzing

23. For each system shown below, indicate the name of the system, and at least one of its functions.

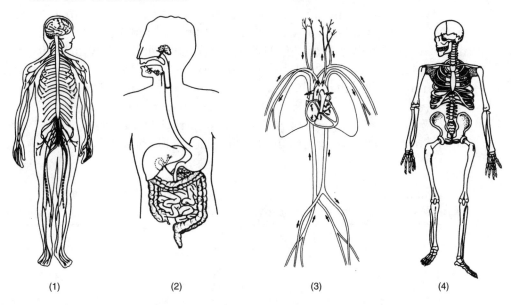

(1)   (2)   (3)   (4)

24. The diagram below shows how nutrients are transported to and from the leaves of a plant. Which human body system performs this function?

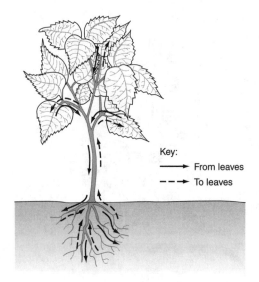

Key:
→ From leaves
- - -→ To leaves

Respiration, Excretion, and Reproduction

The Respiratory System

The circulatory system gets the oxygen it provides to the cells from the *respiratory system*. Cells use this oxygen in the process of **cellular respiration**, in which nutrients from digested food combine with the oxygen to release energy and produce the waste materials carbon dioxide and water. This chemical process takes place in all body cells.

The *respiratory system*, illustrated in Figure 4-11, brings oxygen from the air to the blood, and returns carbon dioxide from the blood to the air. This process is called **respiration**.

When you breathe in *(inhale)*, air enters the nose or mouth and passes through the **trachea**, or windpipe. The trachea branches off to each lung through tubes called **bronchi**. The lungs contain millions of tiny *air sacs*, surrounded by capillaries. Here, respiratory gases are exchanged—oxygen enters the blood while carbon dioxide leaves the blood and is breathed out *(exhaled)*.

The oxygen that enters the blood is carried to the cells of the body, where an exchange of gases again takes place. This time, oxygen leaves the

Figure 4-11. The human respiratory system.

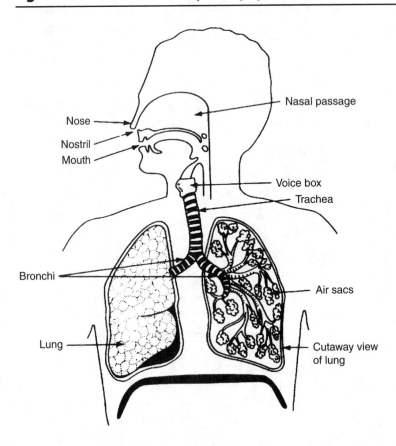

blood and enters the cells, while carbon dioxide leaves the cells and goes into the blood. The carbon dioxide is returned to the lungs to be exhaled. This process is repeated constantly.

The Excretory System

The activities of the body's cells produce waste materials that must be removed. These wastes are removed from the blood, and, eventually, from the body, by the *excretory system*. The excretory system consists of the *lungs, skin, kidneys,* and *liver.*

1. The **lungs** *rid the body of* the waste products carbon dioxide and water vapor each time you exhale.
2. The **skin** gets rid of wastes when you perspire. Microscopic sweat glands deep in the skin excrete *perspiration,* a liquid waste consisting mostly of water and salts. Perspiration leaves the body through the *pores,* which are tiny openings in the surface of the skin (Figure 4-12).
3. The two **kidneys** (Figure 4-13) help to maintain the proper balance of water and minerals in the body. As blood flows through the kidneys, excess water, salts, urea, and other wastes are removed from the blood. These substances make up a fluid called *urine.* Urine is sent through a tube from each kidney to the *bladder,* where it is stored until excreted from the body.
4. The **liver** produces *urea,* a waste resulting from the breakdown of proteins. Urea is taken by the blood to the kidneys and expelled from the body in urine. The liver also removes harmful substances from the blood.

Figure 4-12. *Sweat glands in the skin expel wastes from the body through pores.*

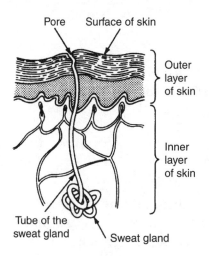

Figure 4-13. *The human urinary system, part of the excretory system.*

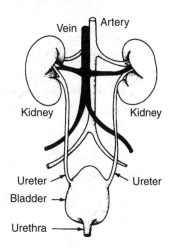

Interpreting a Diagram

The diagram below is a *schematic* representation of the circulatory system. In other words, it is not meant to be a realistic drawing of body parts, but only to show the basic scheme of the system—the relationships among its parts and the sequence of events that occur in the system.

Scheme of blood flow

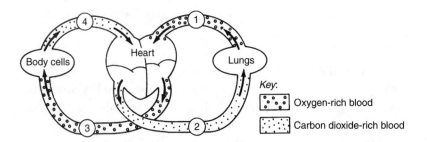

The circulation of blood is vital to the process of respiration, since the blood carries fresh oxygen to the cells of the body and returns carbon dioxide to the lungs to be expelled.

As you have learned, arteries are blood vessels that carry blood away from the heart. Which blood vessels in the diagram are arteries? The arrows indicate that blood vessels 2 and 3 carry blood away from the heart, so they are arteries. Blood vessels 1 and 4, which return blood to the heart, are veins. Study the diagram and then answer the following questions.

QUESTIONS

1. Blood rich in oxygen is found in blood vessels
   (1) 1 and 2   (2) 2 and 3   (3) 1 and 3   (4) 2 and 4
2. Compared with blood vessel 1, the amount of carbon dioxide in blood vessel 2 is
   (1) greater   (2) less   (3) the same
3. Which statement is true?
   (1) All arteries carry oxygen-rich blood.
   (2) All veins carry oxygen-rich blood.
   (3) Arteries from the heart to the lungs carry oxygen-rich blood.
   (4) Veins from the lungs to the heart carry oxygen-rich blood.

Metabolism

Your body carries out an amazing number of chemical reactions that keep you alive. These reactions put things together and break them apart. For example, you take in food, break it down, and use the energy in it; you build and repair tissues; you store fat. *Metabolism* is the sum of all the chemical reactions that take place in the body.

Metabolism can be influenced by hormones, exercise, diet, and aging. Carbohydrate metabolism involves the hormone insulin. People who have diabetes either do not produce insulin or their bodies do not respond to it; therefore, they do not metabolize carbohydrates properly. Exercise increases metabolism because cells need more oxygen. When you eat too much, your metabolism stores the excess as fat. As you reach old age, your metabolism slows.

The Reproductive System

The job of the reproductive system is the production of offspring. There are two human reproductive systems, male and female, as shown in Figure 4-14.

1. *Male.* The male reproductive system consists of the *testes, penis,* and *sperm ducts.* The **testes** produce *sperm cells,* the male sex cells. During reproduction, these cells pass through tubes called **sperm ducts,** where they mix with a fluid to form *semen.* The semen is delivered through the *penis* into the female's reproductive system.
2. *Female.* Making up the female reproductive system are the *ovaries, oviducts, uterus, vagina, and mammary glands.* The **ovaries** produce *egg cells,* the female reproductive cells. Once a month, an egg cell leaves an ovary and travels through one of the **oviducts** to the **uterus,** or womb. If sperm cells are present in the oviduct, fertilization may take place.

After fertilization has occurred, the fertilized egg attaches itself to the inner wall of the uterus. There it develops into a new offspring over a period of about nine months. At the end of this time birth takes place, and the offspring emerges through the **vagina,** or birth canal. The newborn baby may be fed milk produced by the mother's **mammary glands,** or breasts.

Figure 4-14. *The human reproductive systems: male (left) and female (right).*

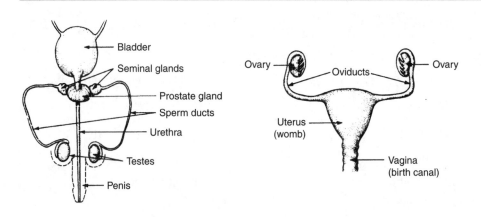

Multiple Choice

25. The process by which energy is released from nutrients is called

   (1) cellular respiration   (3) digestion

   (2) excretion   (4) circulation

26. Where does cellular respiration take place?

   (1) in the blood only   (3) in the heart only

   (2) in the lungs only   (4) in all body cells

27. The respiratory system includes the

   (1) heart, liver, lungs   (3) stomach, esophagus, liver

   (2) lungs, trachea, nose   (4) heart, arteries, veins

28. Which represents the correct order in which oxygen enters the body?

   (1) nose, trachea, bronchi, lungs

   (2) bronchi, nose, trachea, lungs

   (3) lungs, bronchi, trachea, nose

   (4) nose, bronchi, trachea, lungs

29. The exchange of gases between the air and the blood takes place in the

   (1) nose   (2) trachea   (3) bronchi   (4) lungs

30. At each body cell,

   (1) carbon dioxide enters the blood, and oxygen leaves the blood

   (2) both carbon dioxide and oxygen enter the blood

   (3) both carbon dioxide and oxygen leave the blood

   (4) oxygen enters the blood and carbon dioxide leaves the blood

31. Which organ belongs to both the excretory system and the respiratory system?

   (1) heart   (2) kidney   (3) lung   (4) liver

32. The excretory system includes the

   (1) kidneys, liver, lungs   (3) stomach, esophagus, liver

   (2) lungs, trachea, nose   (4) heart, arteries, veins

33. Which of the following helps remove wastes from the body?

   (1) the skull   (3) the spinal cord

   (2) the skin   (4) the stomach

Thinking and Analyzing

Use the diagrams below to answer questions 34 and 35.

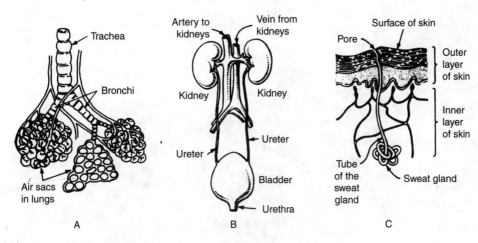

34. To which system do all three structures belong?

    (1) excretory   (2) circulatory   (3) respiratory   (4) skeletal

35. Which diagram may also represent part of the respiratory system?

    (1) A   (2) B   (3) C   (4) all of these

36. Male is to sperm, as female is to

    (1) egg   (2) testes   (3) oviduct   (4) uterus

37. After jogging in place for two minutes (see the Process Skill: "Doing an Experiment" on page 72) Pauline and Jerry noticed that they were out of breath. Explain why their bodies needed more oxygen while exercising?

Chapter *5*
Earth's Surface

Points to Remember

▷ Physical and chemical properties, such as hardness, streak color, cleavage, and reaction to acid are used in mineral identification.

▷ Rocks are composed of one or more minerals. The three types of rocks—igneous, sedimentary, and metamorphic—are classified according to their method of formation. The rock cycle demonstrates how various processes can transform rocks from one type to another.

▷ Fossils are the remains or traces of organisms that lived long ago. They are used to determine past environments and climate. Fossils are usually found in sedimentary rocks.

▷ The study of rock formations provides clues to the history of crustal activity in an area.

▷ Topographic maps are a graphic representation of Earth's surface. Contour lines are used to show the shape of the land.

▷ Earth has a grid system that consists of west-east lines of latitude and north-south lines of longitude. Using a compass and a topographic map, direction points can be accurately determined.

Surface Materials

Bedrock and Soil

Earth's rocky outer layer is called the **crust**. The surface of the crust consists of bedrock, rock fragments, and soil, as shown in Figure 5-1.

Bedrock is the solid rock portion of the crust. Bedrock that becomes exposed at Earth's surface is called an *outcrop*. Rock fragments are pieces of broken-up bedrock. They can range in size from giant boulders to tiny grains of sand.

Soil is a mixture of small rock fragments and *organic matter* (materials produced by living things, such as decaying leaves and animal wastes). Water and air are also important parts of soil. Soil and fragments of rock make up most of Earth's surface, with the bedrock hidden underneath.

Minerals

Rocks are composed of **minerals**, which are naturally occurring solid substances made of inorganic (nonliving) material. Many minerals are found in Earth's crust, but only a few rock-forming minerals make up most of the crust. Feldspar is the most abundant mineral. Some other common minerals are quartz, mica, and calcite. Minerals have certain *physical* and *chemical properties* by which they can be identified.

1. **Physical properties** of minerals include streak color, hardness, luster, cleavage, and color.

Streak color is the color of the powdered form of the mineral. It is obtained by scratching the mineral on the unglazed portion of a porcelain tile. Some minerals have no streak color.

Hardness is the resistance of a mineral to being scratched. Minerals are assigned a number between 1 and 10 to indicate their hardness, with 1 being

Figure 5-1. *Earth's surface consists of bedrock, rock fragments, and soil.*

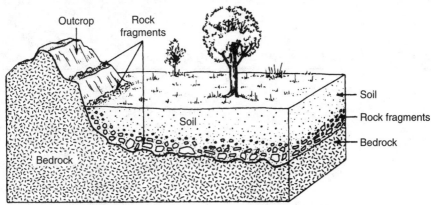

Table 5-1. *Hardness Scale of Minerals*

Mineral	Hardness	Mineral	Hardness
Talc	1	Feldspar	6
Gypsum	2	Quartz	7
Calcite	3	Topaz	8
Fluorite	4	Corundum	9
Apatite	5	Diamond	10

the softest and 10 the hardest. The hardness scale shown in Table 5-1 lists the minerals used as reference points. A mineral can be scratched only by another mineral with a higher number on the hardness scale.

Luster refers to how a mineral looks when it reflects light. A mineral can look metallic, glassy, greasy, or earthy.

Cleavage is a mineral's tendency to break along smooth, flat surfaces. The number and direction of these surfaces are clues to a mineral's identity. Cleavage often causes a mineral to break into characteristic shapes, as shown in Figure 5-2. Not all minerals have definite cleavage; some fracture unevenly when broken.

Color is not always a reliable guide to a mineral's identity. Various samples of the same mineral may have different colors. On the other hand, samples of different minerals may share the same color. Color is best used with other properties to identify a mineral.

2. Minerals also have ***chemical properties***, such as how they react with an acid. For example, calcite, the chief mineral in limestone and marble, fizzes when hydrochloric acid is placed on it. The fizzing is caused by a chemical reaction between the calcite and the acid in which bubbles of carbon dioxide gas are given off. Most chemical tests are difficult to administer in the classroom.

Rocks

The ***rocks*** that form Earth's crust are natural, stony materials composed of one or more minerals. Like minerals, rocks are identified by their physical

Figure 5-2. *Cleavage in* (left) *mica;* (right) *galena.*

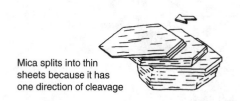
Mica splits into thin sheets because it has one direction of cleavage

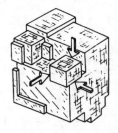
Galena breaks into cube-shaped pieces because it has three directions of cleavage at right angles

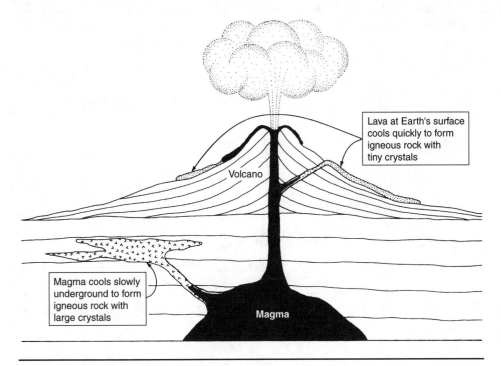

Figure 5-3. *Formation of igneous rocks.*

and chemical properties. Rocks are classified into three groups—igneous, sedimentary, and metamorphic—depending on how they are formed.

1. ***Igneous rocks*** are produced by the cooling and hardening of hot, liquid rock. This melted rock material is called *magma* when underground and *lava* when it pours onto Earth's surface. Extrusive volcanic features are produced when volcanic rock forms on Earth's surface. Intrusive features form underground. Different igneous rocks are generally identified by their color and by the size of the mineral grains (crystals) they contain.

   Igneous rocks that form from rapid cooling of lava, called *volcanic rocks*, contain tiny crystals. Basalt is a dark-colored volcanic rock composed of crystals too small to be seen with the unaided eye.

   Igneous rocks that form underground by slow cooling of magma develop large crystals. Granite is a light-colored igneous rock that contains large, easily visible mineral grains. Figure 5-3 shows processes that produce igneous rocks.

   Magma cools at different rates depending on its depth below Earth's surface. The closer the magma is to the surface, the smaller the grain (crystal) size of the minerals formed in the rock. In fact, lava on the Earth's surface may cool so rapidly that obsidian, a glassy substance forms, not individual grains.

   Carefully study Figure 5-4 on page 86, showing the relationships among common igneous rocks, grain size, rate of cooling, and environment in which magma solidified. Fill in the missing data in a copy of the table shown below the Figure.

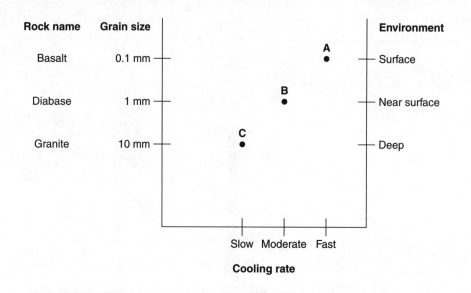

Figure 5-4. *Grain size of igneous rocks depends on the rate and depth at which magma cools.*

Grain Size	Environment	Cooling Rate	Letter in Diagram	Igneous Rock
0.1 mm				
	Near surface		B	
		Slow		

2. *Sedimentary rocks* form from particles called *sediments* that pile up in layers. These sediments may be small rock fragments or seashells. Sedimentary rocks usually form underwater. For example, when a stream carrying particles of sediment empties into an ocean or lake, the particles settle to the bottom in layers. Eventually, these layers harden into sedimentary rock (see Figure 5-5). Table 5-2

Figure 5-5. *Sedimentary rocks form in layers.*

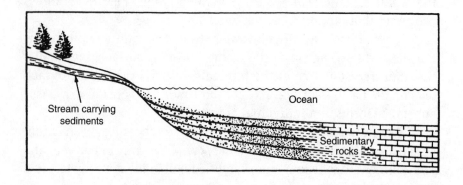

Table 5-2. Common Sedimentary Rocks

Rock Name	Type of Sediment	Place of Formation
Sandstone	Sand grains	Shallow waters near a shore pounded by waves
Shale	Clay particles	Deep, calm ocean waters; lake bottoms
Limestone	Tiny seashells	Warm, shallow seas

above lists some common sedimentary rocks, what they are made of, and where they form.

When entering the ocean, a river or stream loses energy and drops the sediments it is carrying. It drops the largest sediments first and, as the water flow continues to slow, it drops smaller and smaller particles. After sediments accumulate for millions of years, they become buried deep in the earth and harden into sedimentary rocks.

Figure 5-6 shows the accumulation of sediments as a river enters the ocean.

Sedimentary Rock Characteristics

Sedimentary Rock	Description	Size of Particles	Formation Environment
Sandstone	Cemented grains of sand	0.06–2.0 mm	_____
Conglomerate	Visible cemented rounded pebbles	> 2.0 mm	_____
Shale	Sheets of tightly compact clay particles	< 0.004 mm	_____
Siltstone	Powdery grains of cemented silt particles	0.004–0.06 mm	_____

Using Figure 5-6, which shows depositing locations of sediment, select the letter that best shows the formation environment of each of the sedimentary rocks listed. Place the letter in the formation environment column in a copy of the table shown above.

3. **Metamorphic rocks** are produced when either igneous or sedimentary rocks undergo a change in form caused by heat, pressure, or both. This can take place when magma heats rocks it comes in

Figure 5-6. Sediments accumulate on the ocean floor.

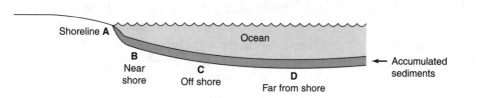

contact with or when forces deep underground squeeze rocks for long periods of time. The high temperatures and pressures thus created change the appearance and mineral composition of the rocks, transforming them into metamorphic rocks.

Marble and slate are metamorphic rocks formed from the sedimentary rocks limestone and shale, respectively. Gneiss (pronounced "nice") is a metamorphic rock that can be produced from granite, an igneous rock.

 ## Laboratory Skill

Identifying Minerals Using a Flowchart

Each mineral has a set of specific physical and chemical properties that can be used to identify it. The properties of a mineral are best obtained from a fresh clean surface; therefore, breaking the specimen may be necessary. Mineral identification tables organize these properties for ease of identification.

The following steps describe the process of identifying an unknown mineral specimen.

1. Streak color: Scratch the mineral specimen on the unglazed side of a porcelain tile. If a streak appears on the tile, determine the color of the streak. If no streak appears on the tile, the specimen does not have a streak color.

2. Color: Determine the visible color of the specimen. Any shade of black, dark green, brown, red, or blue is considered dark colored; otherwise the specimen is light colored.

3. Hardness: Using simple tools such as a glass scratch plate and a carpenter's nail (hardness of each between 5–6) determine if the mineral is hard (can scratch the glass plate), or soft (can be scratched by the nail.)

4. Cleavage or Fracture: This is best determined by breaking the specimen and observing how it breaks. Generally, if the specimen shows flat surfaces, corners, or edges it has cleavage; otherwise it has fracture.

5. After you determine the properties of the mineral use the proper table below to identify your specimen. If the specimen has no streak, use Table A. If it has a streak, use Table B. To use Table A, determine whether the specimen is light- or dark-colored; hard or soft; has cleavage or fracture; and where it fits in the color and properties listed. To use Table B, first determine the streak color. Then decide whether it is hard or soft, has cleavage or fracture, and if it fits into color and properties listed.

6. Matching the specimen with known minerals, pictures of minerals, and more extensive descriptions can help assure final identification.

(Continued)

TABLE A. NO STREAK

Color	Hard/Soft	Cleavage/Fracture	Common Colors/Properties	Name
Light Colored	Hard	Cleavage	Gray, white, flesh colored	Feldspar
		Fracture	White, looks waxy	Milky quartz
			Pink, looks glassy to waxy	Rose quartz
	Soft	Cleavage	Clear, salty taste	Halite
			White, soft, scratch with fingernail	Gypsum
			Very soft, soapy feel	Talc
			Colorless, thin sheets peel easily	Muscovite mica
			White to gray, fizzes in weak HCl acid	Calcite
		Fracture	Tan, earthy	Bauxite
Dark Colored	Hard	Cleavage	Black, elongated grains, hardness 5–6	Hornblende
		Fracture	Red, brown, green, looks glassy	Garnet
			Red, brown, yellow, dull or waxy	Jasper
			Black, gray, dull or waxy	Flint
			Gray to black, glassy to waxy	Smoky quartz
	Soft	Cleavage	Black, brown, thin sheets peel easily	Biotite mica
		Fracture	Green to black, soapy feel	Serpentine

TABLE B. STREAK COLOR

Streak Color	Hard/Soft	Cleavage/Fracture	Common Colors/Properties	Name
Black/Dark Green/or Gray Streak	Hard	Cleavage	No common minerals	
		Fracture	Black, magnetic (magnet sticks to it)	Magnetite
			Brassy yellow, looks metallic	Pyrite
	Soft	Cleavage	Metallic silver, cubes	Galena
		Fracture	Black to gray, marks paper, feels greasy	Graphite
Red/Brown Streak	Hard	Cleavage	No common minerals	
		Fracture	Red to brown, hardness 5–6	Hematite
	Soft	Cleavage	Yellow, brown, looks glassy or waxy	Sphalerite
		Fracture	Yellow-brown, earthy luster	Limonite
			Silver-gray, tiny flakes, looks metallic	Specularite
Green Streak	Soft	Fracture	Bright green, looks earthy, with Azurite	Malachite
Blue Streak	Soft	Fracture	Bright blue, looks earthy, with Malachite	Azurite

(Continued)

1. You can distinguish between halite and calcite by
   (1) using a magnet
   (2) placing a drop of HCl on each mineral
   (3) observing cleavage
   (4) determining streak color.
2. An unknown mineral sample scratches a glass plate, but cannot be scratched by a carpenter's nail. The mineral
   (1) is considered hard
   (3) has an hardness between 5 and 6
   (2) is considered soft
   (4) may be graphite
3. The property of the mineral in the diagram that is most easily recognized is _____.

4. Mary has an unknown mineral sample with a fresh surface that she wants to identify. She observes the following characteristics of the mineral

   • It is a dark red-brown color.
   • It produces a reddish streak color.
   • It scratches a glass plate.
   • It does not have corners, flat surfaces, or straight edges.

   What is the name of the mineral?
5. Using a computer database, design a chart that will allow you to enter the properties needed to identify five unknown specimens as described in the Process Skill above.

Process Skill

How Do Rocks Change from One Type to Another Type?

The three types of rocks—igneous, sedimentary, and metamorphic—can be transformed into a new type of rock by natural processes. Weathering and erosion of igneous, sedimentary or metamorphic rocks produce sediments that can be buried and cemented into new sedimentary rocks. Heat and/or pressure deep within Earth's crust can affect previously existing rocks, transforming them into new met-

(Continued)

amorphic rocks. Heating associated with volcanism can melt old rock into a liquid state that solidifies into new igneous rocks. Basically, over long periods of time, old rock material is recycled into new rock material many times. All these changes and processes make up the rock cycle shown in the diagram below. Study the diagram and answer the questions that follow.

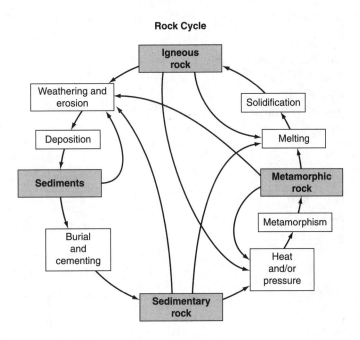

Rock Cycle

QUESTIONS

1. The processes necessary to change a metamorphic rock into a sedimentary rock are
   (1) weathering and erosion, deposition, burial and cementation
   (2) melting and solidification
   (3) heat and pressure
   (4) none of the above
2. Which of these sequences *cannot happen*?
   (1) igneous rock → heat and/or pressure → metamorphic rock
   (2) igneous rock → weathering and erosion → deposition → burial → sedimentary rock
   (3) igneous rock → melting and solidification → igneous rock
   (4) all three sequences are possible
3. Using the rock cycle chart, how many years does it take an igneous rock to become a sedimentary rock?
   (1) 100 years   (3) 1,000,000 years
   (2) 500 years   (4) cannot be determined
4. What type of rock is produced by the solidification of a liquid rock mixture.
5. Give two possible sequences of processes that could change sedimentary rocks into igneous rocks.

Multiple Choice

1. Bedrock exposed at Earth's surface is called

   (1) a mountain
   (3) an outcrop

   (2) a boulder
   (4) a rock fragment

2. The graph indicates the hardness of six minerals. Which mineral is hard enough to scratch fluorite, but will not scratch garnet?

   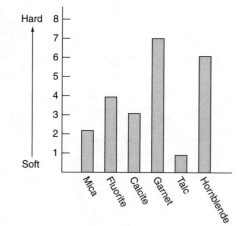

   (1) mica

   (2) calcite

   (3) hornblende

   (4) talc

3. A *chemical* property that would help to identify a mineral is

   (1) luster
   (3) reaction to acid

   (2) hardness
   (4) cleavage

4. Rocks that form from layers of small particles are called

   (1) metamorphic rocks
   (3) igneous rocks

   (2) sedimentary rocks
   (4) volcanic rocks

5. *Schist* is a metamorphic rock. This means it was formed by

   (1) cooling and hardening of magma

   (2) great heat or pressure, or both

   (3) buildup of sand grains

   (4) buildup of clay particles

6. What process was involved in forming the mountain shown below?

   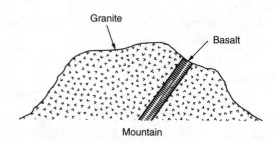

   (1) cooling of magma to form igneous rock

   (2) buildup of sediments in shallow water

(3) deep underground pressure

(4) buildup of sediments in deep water

7. Granite has large mineral grains because it is formed by

(1) slow cooling of magma

(2) cementation of large rock fragments

(3) rapid cooling of lava

(4) high pressures

8. Casey finds a rounded rock in a streambed in New York State. He knows the rock was part of a much larger rock structure somewhere upstream. Where is the rock in the rock cycle?

(1) A

(2) B

(3) C

(4) D

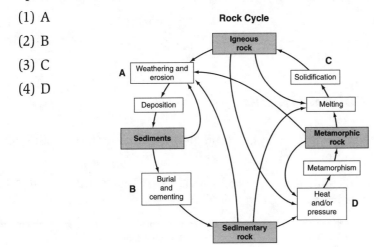

Rock Cycle

After reviewing the characteristics of four similar looking mineral specimens, use the chart below to distinguish among them and identify each. Then answer questions 9–12.

Specimen 1: Does not react to acid, white color, and can easily be scratched by specimen 3

Specimen 2: Reacts to acid, white color, and can easily be scratched by specimen 3

Specimen 3: Does not react to acid, white color, none of the other specimens will scratch it

Specimen 4: Does not react to acid, white color, can only be scratched by specimen 3

Mineral Identification Chart

Mineral Name	*Hardness*	*Acid Test*	*Common Color*
Quartz	7	No reaction	White
Calcite	3	Reaction	White
Gypsum	2	No reaction	White
Feldspar	6	No reaction	White

9. Specimen 1 is most likely
   - (1) quartz
   - (2) calcite
   - (3) gypsum
   - (4) feldspar

10. Specimen 2 is most likely
    - (1) quartz
    - (2) calcite
    - (3) gypsum
    - (4) feldspar

11. Specimen 3 is most likely
    - (1) quartz
    - (2) calcite
    - (3) gypsum
    - (4) feldspar

12. Specimen 4 is most likely
    - (1) quartz
    - (2) calcite
    - (3) gypsum
    - (4) feldspar

Thinking and Analyzing

13. Describe the steps that you would perform to identify an unknown mineral specimen.

14. The following table lists some properties of two common minerals, quartz and calcite.

	Common Color	Hardness	Cleavage/Fracture	Other:
Quartz	White/clear	7	Fracture	
Calcite	White	3	Cleavage	Fizzes with HCl

Given a sample of each mineral, how might you determine which is quartz and which is calcite?

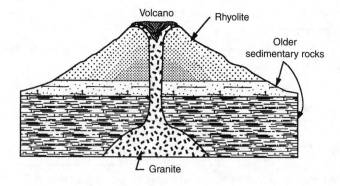

15. What type of rock is produced by volcanic activity?

16. Rhyolite and granite have the same chemical composition and are formed from the same liquid rock solution. Why does rhyolite have a grain size of less than 1 mm while granite has a grain size of greater than 1 mm?

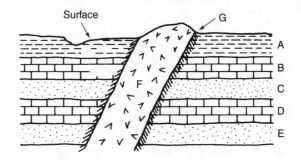

17. Rock layers A, B, C, D, and E are sedimentary rocks. Letter F represents an intrusion of magma that solidified into granite. What type of rock would you expect to find at G, the contact zone between the granite and the sedimentary rock?

Earth History

Interpreting Rocks

Scientists have pieced together much of Earth's history by studying rocks all over the world. The rocks in an area contain much information about that area's past. For example, the presence of sedimentary rocks indicates that an area was once covered by water. Fossils in sedimentary rocks tell of past life forms and the environments in which they lived.

Scientists can interpret clues in rocks that reveal the order in which they were formed. Horizontally layered sedimentary rocks are easiest to interpret. The bottom layers were laid down first and are therefore the oldest, while the youngest layers are at the top (Figure 5-7).

This simple situation is often complicated by later events. Folding and faulting of rock layers and intrusions of igneous rock material make it difficult to tell which rocks are the oldest and which are the youngest (Figure 5-8 on page 96). However, all these features are clues to events in Earth's past and the order in which they happened.

Figure 5-7. *In a stack of sedimentary rocks, the oldest layers are at the bottom and the youngest at the top.*

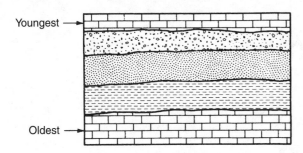

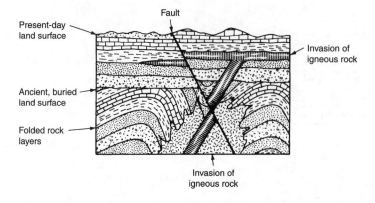

Figure 5-8. A rock sequence with a complex history of folding, erosion, volcanism, and faulting.

Fossils

Fossils are the remains or traces of organisms that lived long ago. Fossils are formed when a dead plant or animal, or some trace, like a footprint in mud, is covered by sediment that later hardens into rock. Almost all fossils are found in sedimentary rock. Figure 5-9 shows several types of fossils.

Scientists have learned much about Earth's past by studying fossils. Fossil evidence has helped scientists to trace the evolution of life from simple ancient organisms to complex present-day life forms. Fossils also provide clues to ancient environments. For example, corals live only in warm, sunlight-rich waters. Finding fossil corals in central New York suggests that the area was once covered by a warm, shallow sea.

Fossils can sometimes be used to match up rock layers that are far apart. Finding the same group of fossils in rock layers at separate locations indicates that those layers formed in the same time period (Figure 5-10).

Dating Rocks

The relationships between rock layers, fossils, folds, faults, and intrusions of igneous rock can indicate the order of events in Earth's past. However, they cannot reveal the actual age of the rocks. To determine the age of rocks, scientists use a technique called *radioactive dating*.

Figure 5-9. Examples of some fossil types.

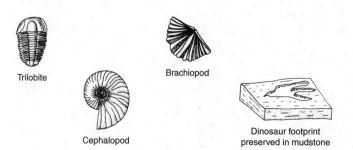

Trilobite

Brachiopod

Cephalopod

Dinosaur footprint preserved in mudstone

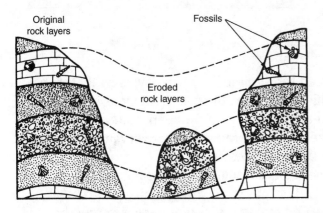

Figure 5-10. *Fossils can be used to match up distant rock layers.*

Most rocks contain small amounts of radioactive substances that change (decay) into nonradioactive substances at a definite rate. For example, radioactive *uranium* changes into lead at a known rate. By measuring and comparing the amounts of uranium and lead in a rock, the age of the rock can be determined.

Using this technique, scientists have been able to assign dates to major events in Earth history, such as periods of mountain building, the formation of oceans, and the appearance of various life forms. Scientists estimate that Earth itself is about four and a half billion years old.

Process Skill

Explaining a Relationship; Interpreting Diagrams

Many events from Earth's past are recorded in rocks. By examining features in rocks, such as folds, faults, and invasions of igneous rock, the sequence of events that produced present-day rock structures can often be sorted out.

For example, observe the rock structure in Diagram 1 below. Notice that the igneous rock *cuts across* the pattern of sedimentary rock layers. For this to happen, the sedimentary layers must have already been in place. Therefore, the sedimentary rocks were formed first.

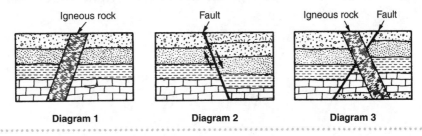

(Continued)

In Diagram 2, the fault has shifted the sedimentary rock layers so that they do not match up across the fault. This means that the sedimentary layers were formed first. The direction in which matching layers have been shifted indicates that the rocks on the right side of the fault have moved downward in relation to the rocks on the left side, as shown by the arrows.

Diagram 3 is more complex. The fault has shifted the sedimentary layers, so the sedimentary rocks were formed before the fault. The igneous rock cuts across the sedimentary layers, so the sedimentary rocks were also formed before the igneous rock. But the fault has *not* shifted the igneous rock, so the igneous rock must have formed *after* the fault. The order of formation here is: sedimentary rocks, fault, igneous rock. Examine Diagram 4 and answer the questions below.

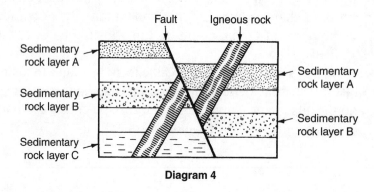

Diagram 4

QUESTIONS

1. Of the events that produced this rock structure, which occurred first?
   (1) formation of igneous rock
   (2) faulting
   (3) formation of sedimentary rock layer *C*
   (4) formation of sedimentary rock layer *A*
2. What is the correct order of formation in this rock structure?
   (1) sedimentary rocks, igneous rock, fault
   (2) sedimentary rocks, fault, igneous rock
   (3) igneous rock, sedimentary rocks, fault
   (4) igneous rock, fault, sedimentary rocks
3. Which set of arrows correctly shows the directions in which rock layers were shifted along the fault in Diagram 4?

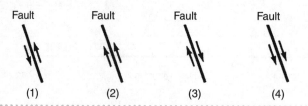

Multiple Choice

18. Scientists can use fossils in rocks to

    (1) match up rock layers at different places

    (2) study the evolution of life

    (3) learn about past environments

    (4) all of the above

19. Valerie identified a rock outcrop near her school as shale, a sedimentary rock. What does this suggest about the region's past?

    (1) Underground volcanic activity once took place.

    (2) Surface volcanic activity once took place.

    (3) The area was once underwater.

    (4) Great heat or pressure once affected the area.

20. By studying folds and faults in rocks, scientists can determine

    (1) the age of a rock in years

    (2) the order of past Earth events

    (3) the depth of the ocean

    (4) if there is life on Mars

21. The diagram shows layers of sediments deposited in a body of water. Which layer was deposited first?

    (1) layer *A*

    (2) layer *B*

    (3) layer *C*

    (4) layer *D*

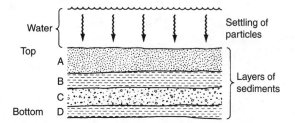

22. Most fossils are found in

    (1) igneous rocks   (3) metamorphic rocks

    (2) sedimentary rocks   (4) volcanic rocks

23. Carol found a coral fossil in a limestone outcrop near her home. Which statement about the area's past is most likely correct?

    (1) The area was once under deep, cold water.

    (2) The area was never underwater.

    (3) The area was once under cold, shallow water.

    (4) The area was once under warm, shallow water.

24. A rock was found containing fossils of three different ancient organisms. The table shows the time ranges in which these organisms lived. The rock must have been formed at a time when all three organisms were living. What is the best estimate of when the rock was formed?

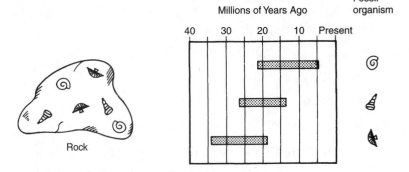

(1) 10 million years ago

(2) 15 million years ago

(3) 20 million years ago

(4) 25 million years ago

25. Scientists use radioactive dating to

(1) determine the age of a rock in years

(2) determine the order of events in Earth history

(3) identify the rock type in a region

(4) learn about past climate and weather

Thinking and Analyzing

Use the cross section diagram below to answer questions 26–29.

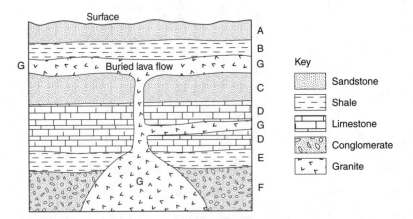

26. Which rock type is the oldest?

27. Describe what event happened after layer C was deposited and before layer B was deposited.

28. What type of rock is G?

29. Describe what the surface was like when layer D was deposited.

Visualizing Earth's Surface

Maps

A **map** is a representation of a portion of Earth's surface. Different types of maps are used to show specific information about Earth's surface and provide useful information for people. Weather maps show weather conditions and provide important information to airplane pilots, boaters, and skiers. Geologic maps show rock formations and provide geologists with information about the location of oil and mineral deposits. Road maps show the location of roads, towns, or cities, and assist in everyday travel for many drivers.

Topographic maps (Figure 5-11) show the physical features of a small section of Earth's surface. These physical features include mountains, valleys, hills, plains, rivers, lakes, etc. Topographic maps are used by engineers when developing roads, by hikers planning a hike, and by anyone interested in knowing the shape of the land.

Latitude and Longitude

Earth is a sphere, and a sphere can be cut horizontally or vertically into circles. Each circle contains 360 degrees of arc. These circles form the basis of latitude and longitude. The distance between the North Pole and the equator represents one-quarter of the distance around Earth, or 90°. If a circle were

Figure 5-11. *A topographic map of an island. The contour interval is 10 meters. Based on the scale, the distance from point x to point y is 1.5 kilometers.*

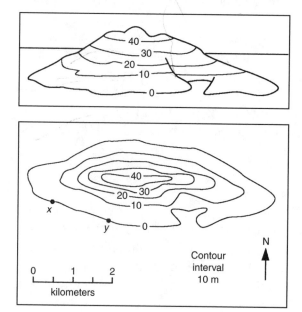

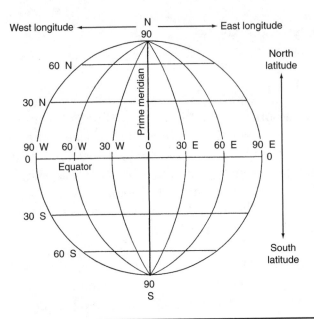

Figure 5-12. *Longitude and latitude lines on a globe.*

drawn around Earth at each degree interval, this would represent a line of **latitude**. The equator is 0° latitude, and as you travel north the lines of latitude increase until you reach the North Pole, which is 90° north latitude. Lines of latitude run west-east, and are measured in degrees north-south. In the Southern Hemisphere, the equator is 0° and the lines of latitude increase until you reach the South Pole, which is 90° south latitude. (Figure 5-12).

Lines of **longitude** run north-south, and are measured in degrees west-east. The 0° longitude line, the prime meridian, is an imaginary line that connects the North and South Poles and passes through Greenwich, England. From the prime meridian, you can travel east one-half the distance around Earth in the east longitude hemisphere to the 180° longitude line, or you can travel west one-half the distance around Earth to the 180° longitude line.

For maps that show large areas, such as continents or oceans, the unit of degrees for both latitude and longitude measurements is sufficient. However, for maps showing smaller features, such as topographic maps, a smaller unit of measure is needed. A degree of arc is divided into 60 smaller units called minutes, and a minute of arc is divided into 60 smaller units called seconds. Minutes and seconds are used here as units of angle measurement, and not as units of time.

Topographic Maps

The shape of the land is shown on topographic maps by the use of **contour lines**, lines connecting points of equal elevation. Contour lines represent land elevations above mean sea level. For ease of reading elevation, every 5th contour line is an index contour line. It is darker, and its elevation is labeled.

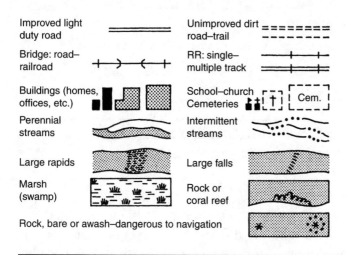

Figure 5-13. *Symbols on topographic maps.*

Topographic map size is determined by the amount of latitude and longitude shown on the map. These maps cover a small area and are measured in minutes rather than degrees. Typical topographic map sizes are 7.5 × 7.5 minutes, 15 × 15 minutes, and 30 × 30 minutes. A 7.5 × 7.5 minute map represents an area of land 7.5 minutes of latitude by 7.5 minutes of longitude.

In addition to showing the shape of the land, topographic maps use colors and symbols to show other features. Blue is used for bodies of water such as lakes and rivers. Features such as roads, buildings, railroad tracks, and cemeteries are black. Green is used to show areas of dense vegetation such as woodlands. Contour lines are brown. Different types of symbols are used on topographic maps to show many natural and constructed features. Figure 5-13 shows a small number of such symbols.

Topographic Map Information

Along the margin of topographic maps there is much information about the map. Some of the information includes:

- Map Name—The name of a topographic map is taken from a major geographic or man-made feature on the map. Typical names may come from a river, mountain, city, landform, etc.
- Year of Production and/or Revision—Year of production indicates the year the map was first published and/or last revised.
- General Location in State—A small black box within an outline of the state indicates the map's location within a state.
- Map Scale Ratio—The map scale ratio is written in numerical form; for example—1:62,500 or 1:24,000. This represents the ratio of map units to Earth's surface units. The ratio of 1:24,000 means 1 cm on the map equals 24,000 cm (240 meters) on Earth, or 1 in. on the map equals 24,000 in. (2000 feet) on Earth.

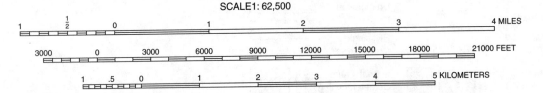

Figure 5-14. *A numerical scale (1:62,500) and three graphic scales (miles, feet, and kilometers) from a topographic map.*

- Distance Scale Graphic Line—The graphic scale is a line divided into segments and labeled with the distance it represents (Figure 5-14).
- Contour Interval—The contour interval is the vertical difference between adjacent contour lines. Typical contour intervals for a topographic map are 10, 20, or 50 feet.
- Difference Between Magnetic North Pole and Geographic North Pole—The difference between magnetic north (MN) and true north (★) is indicated in a diagram showing the angular difference between their directions.
- Latitude and Longitude Labels—The latitude and longitude of a topographic map is labeled in each of the map corners and along each of the four sides of the map at appropriate locations.

Landforms

Landforms are shown by the shape and spacing of contour lines (Figure 5-15). For example:

- Closely spaced contour lines indicate a steep slope, and widely spaced contour lines indicate a gentle slope or nearly flat area.
- Circular contour lines indicate a hill or mountain.
- Contour lines that cross a stream or river make a "V" shape that points upstream.
- U-shaped bends in contour lines indicate a wide, deeply eroded valley.
- Hachure marks (short inward pointing lines) on a contour line indicate a depression.

Profiles

A **profile** is a side view of a landform projected from a straight line on a topographic map. By drawing a profile between two points on a map you get a true representation of the landscape you would encounter if you traveled from one point to the other point. For instance in Figure 5-16, if you were to walk from point M to point N in a straight line you would have to climb over a hill.

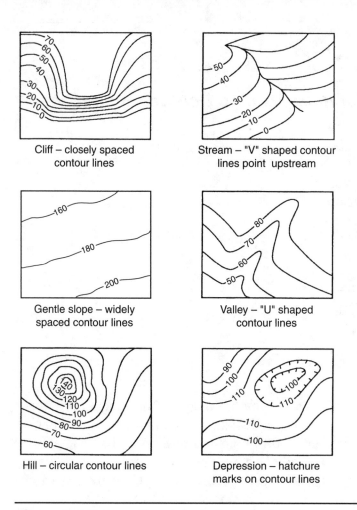

Cliff – closely spaced
contour lines

Stream – "V" shaped contour
lines point upstream

Gentle slope – widely
spaced contour lines

Valley – "U" shaped
contour lines

Hill – circular contour lines

Depression – hatchure
marks on contour lines

Figure 5-15. *From the shape of a set of contour lines, you can recognize various landscapes.*

Figure 5-16. *Making a profile from a contour map.*

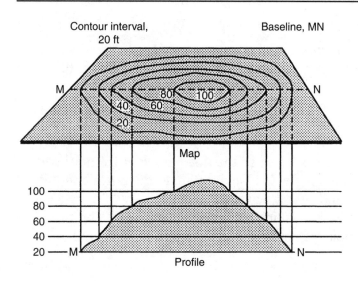

Process Skill

How Is a Compass Used to Determine Direction?

A compass consists of a magnetized metal needle that freely pivots above a circular dial that is labeled with the major geographic direction points. The most commonly labeled points are North, Northeast, East, Southeast, South, Southwest, West, and Northwest. (Diagram 1) However, some compasses divide the circle into 360°—North 0° (or 360°), East 90°, South 180°, and West 270°. Other compasses use 16 or 32 direction points. When allowed to freely swing and rest, the compass needle points to magnetic north (MN).

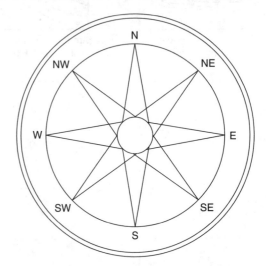

Diagram 1

Earth's iron rich core acts much like a giant bar magnet. The magnetized needle of the compass aligns with Earth's magnetic field and points toward the magnetic north pole. However, the magnetic north pole and true north, or Geographic North Pole (usually indicated by a ★ symbol on a topographic map), are at different locations. In New York State, magnetic north is between 8–15 degrees west of true north (Diagram 2). The difference between magnetic north and true north can be found on the topographic map margin. (Diagram 3) When accuracy of direction is required this difference must be taken into account. Topographic maps are drawn with true north toward the top margin of the map.

To use a compass to determine accurate direction:

1. From a local topographic map, obtain the difference between magnetic north and true north. For this example magnetic north is 15° west of true north.

2. Hold the compass flat and still so that the needle swings freely. The metal needle will eventually settle and point in a direction. This is magnetic north.

3. Rotate the compass and align the directional dial of the compass so that North is located at the end of the needle pointing toward the magnetic north pole.

4. Rotate the compass counterclockwise so the needle points 15° northwest of north. If you do not have degree marking, estimate the angle one-third the distance between N and NW, which is 45° angle.

5. At this position, the needle is pointing toward magnetic north and the compass markings are pointing to the correct geographic directions.

(Continued)

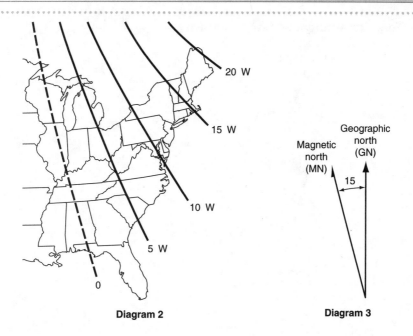

Diagram 2

Diagram 3

QUESTIONS

1. Explain why a compass needle does not point true north to the Geographic North Pole.

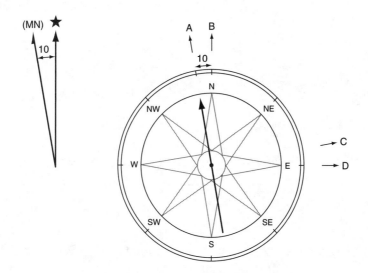

Using the compass diagram and the map symbol showing the difference between magnetic north and true north answer questions 2–4.

2. The arrow labeled A is pointing in what direction?

3. The arrow labeled B is pointing in what direction?

4. Which arrow (C or D) is pointing in a true east direction?

(Continued)

5. In New York State magnetic north is always

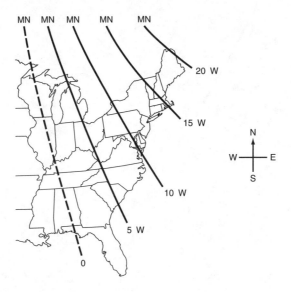

(1) west of Geographic North
(2) east of Geographic North
(3) south of Geographic North
(4) in the same direction as Geographic North

 Review Questions

Multiple Choice

Questions 30–33 refer to the topographic map below.

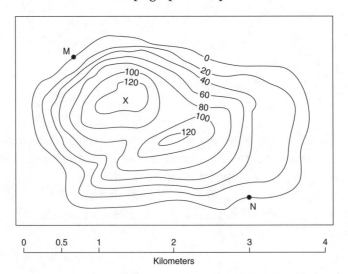

30. The contour interval of the map is

(1) 10 feet   (2) 20 feet   (3) 50 feet   (4) 100 feet

31. The elevation of point X is most likely

(1) 100 feet   (2) 120 feet   (3) 140 feet   (4) 130 feet

32. The distance between points M and N is most nearly

(1) 1 kilometer   (2) 2 kilometers   (3) 3 kilometers   (4) 5 kilometers

33. Which profile best represents the landscape along line M–N?

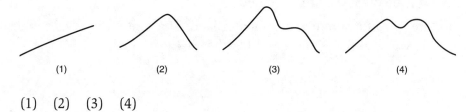

(1)   (2)   (3)   (4)

(1)   (2)   (3)   (4)

34. Using the small topographic map below, which statement is correct?

(1) Point A is higher than point B.

(2) Point B is higher than point A.

(3) Points A and B are at the same elevation.

(4) The relationship between points A and B cannot be determined.

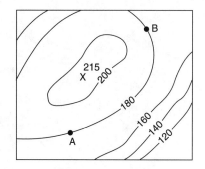

35. Circular contour lines indicate a

(1) hill   (2) valley   (3) depression   (4) ridge

Thinking and Analyzing

The diagram represents a portion of Earth's latitude and longitude grid. Use the diagram to answer questions 36–38.

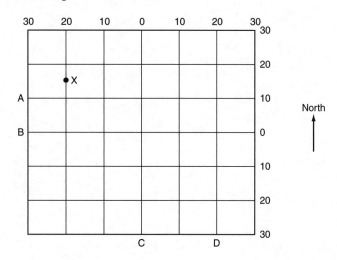

36. Describe line A.

37. What is the latitude and longitude of point X?

38. What line represents the equator?

39. A compass is used to determine

    (1) the direction of true north

    (2) the direction of magnetic north

    (3) the difference between true north and magnetic north

    (4) the direction of the equator

40. Design a topographic map of an island. The contour interval should be 10 feet and the island has a hill that is 45 feet high.

Chapter 6
Forces that Change Earth's Surface

Points to Remember

▷ External forces associated with weathering and erosion wear down Earth's surface. Internal forces associated with faulting and volcanism build up Earth's surface.

▷ Weathering is the breaking down of rocks into smaller pieces. Weathering is caused by physical agents such as ice wedging and chemical agents such as oxygen rusting the iron in rocks.

▷ Erosion is the transport of rock material from one place to another. Running water is the major agent of erosion.

▷ Change in Earth's surface is the result of the interactions of the lithosphere, hydrosphere, and atmosphere over various ranges of time.

▷ Plate tectonics explains how the crust consists of a series of plates that move and interact causing earthquakes and volcanism. Most earthquakes and volcanoes are located on the boundaries of plates.

▷ Major seafloor features are associated with plate tectonics. Mid-ocean ridges are areas where hot rock material comes to Earth's surface and pushes outward in both directions. Trenches are deep, ocean-floor features where one plate slides down under another plate.

▷ Continental drift was initially supported by the fit of the continents, fossil correlation, and rock formations. Today plate tectonics and ocean-floor features strongly support that the continents were once together.

Interacting Earth Systems

Planet Earth can be thought of as consisting of three different spheres: a rock sphere, or **lithosphere**; a water sphere, or **hydrosphere**; and a gaseous sphere, or **atmosphere** (Figure 6-1). On Earth's surface, these three spheres come into contact with one another, thereby affecting one another through the interaction of energy and matter.

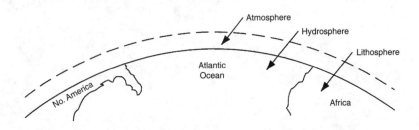

Figure 6-1. *Earth consists of three "spheres": the lithosphere, hydrosphere, and atmosphere.*

Energy is constantly being exchanged among the lithosphere, atmosphere, and hydrosphere. Energy effects can be observed in processes on Earth's surface. Table 6-1 gives examples of energy exchange between Earth's spheres.

Table 6-1. *Examples of Energy Exchange Between Earth's Spheres*

Atmosphere ⟶ Lithosphere

Physical and chemical weathering by atmospheric gases and moisture cause rocks to crumble.

Atmosphere ⟶ Hydrosphere

Waves are produced by wind blowing across water surfaces. The stronger the wind, the larger the waves.

Lithosphere ⟶ Hydrosphere

Tsunami waves are produced by volcanic and earthquake activity on the ocean floor. Strong vibrations of the seafloor are transferred to the water, causing large waves to form.

Hydrosphere ⟶ Lithosphere

Ocean waves breaking along beaches transport sand particles, causing coastline erosion.

Hydrosphere ⟶ Atmosphere

Climate is affected by warm ocean currents traveling north and warming the land they contact, and by cold ocean currents traveling south and cooling the land they contact.

Lithosphere ⟶ Atmosphere

Volcanic activity sends ash particles high into the atmosphere. These particles decrease the sun's radiation reaching Earth's surface, producing cooler temperatures.

External Forces

Various internal and external forces are constantly at work shaping and changing Earth's surface. *Internal forces* tend to push the land up above sea level, and *external forces* tend to wear the land down to sea level. Figure 6-2 illustrates these forces and their effects on Earth's surface features.

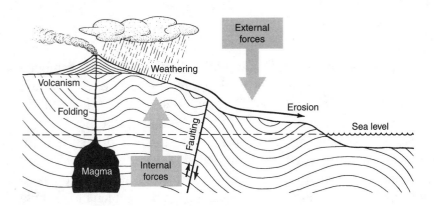

Figure 6-2. *Earth's surface is shaped by the interaction of internal and external forces.*

Weathering

External forces include the processes of weathering and erosion. Together, these processes wear down Earth's surface.

Weathering is the breaking down of rocks into smaller pieces, primarily by agents of weathering: rain, ice, and atmospheric gases. Both physical and chemical agents can cause weathering. In *physical weathering*, rocks are broken into smaller fragments by physical agents. For example, when water seeps into cracks in a rock and freezes, the water expands, as shown in Figure 6-3. The roots of plants growing in cracks can also split rocks apart.

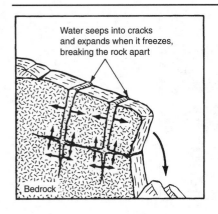

Figure 6-3. *Physical weathering caused by water freezing in rock cracks.*

Chemical weathering is the breaking down of rocks through changes in their chemical makeup. These changes take place when rocks are exposed to air or water. For instance, when rainwater combines with carbon dioxide in the air, a weak acid is formed that dissolves certain minerals in rocks and causes the rocks to fall apart. Also, when oxygen and water react chemically with iron-bearing minerals in a rock, the iron is changed into rust, which crumbles away easily.

By breaking down rocks into smaller fragments, the processes of weathering assist in the formation of soil.

Erosion

Erosion is the process whereby rock material at Earth's surface is removed and carried away. Erosion requires a moving force, such as flowing water, which can carry along rock particles. This can be seen after a heavy rain, when streams turn a muddy brown from the rock material in the water.

Gravity and *water* play important roles in erosion. Gravity is the main force that moves water and rock downhill. Flowing water is very powerful; more rock material is eroded by running water than by all other forces of erosion combined. The Grand Canyon in Arizona is a spectacular example of erosion caused by running water (Figure 6-4)

Groundwater and glaciers are other forces that cause erosion. *Groundwater* forms from rain or snowmelt that filters into the soil. As groundwater seeps through cracks in the bedrock, the water dissolves rock material and carries it away. Eventually, this action may create large underground caves.

Glaciers are masses of ice that form in places where more snow falls in winter than melts in summer, such as in a high mountain valley. The snow that does not melt piles up over the years, and its increasing weight changes the bottom layers into ice. Gravity causes the ice to flow downhill, like a river in slow motion. As a glacier creeps along, it grinds up and removes rock material from the land surface.

Figure 6-4. *Erosion by running water carved the Grand Canyon, a gorge more than a mile deep.*

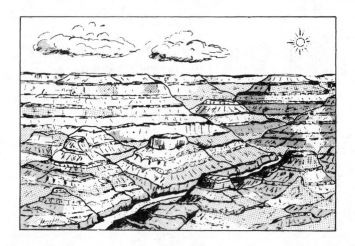

Wind can also act as a force of erosion. In dry desert areas, sand grains blown along by the wind scrape and scour rock outcrops, slowly carving them into unusual shapes.

The forces of erosion are constantly at work, moving rock material from the continents into the ocean basins.

Process Skill

Predicting an Experimental Result

Rocks in a stream constantly knock and scrape against each other and against the streambed as they are carried along by the flowing water. The longer the rocks are in the stream, the more they tumble about and strike one another. To simulate this action and study its effects, a student carried out the following experiment.

Twenty-five marble chips and a liter of water were placed in a large coffee can marked *A*. The can was then covered with a lid and shaken for 30 minutes. Then 25 marble chips and a liter of water were placed in a second can, marked *B*, and covered. This can was shaken for 120 minutes. The illustration below shows the materials used in the experiment. Keep in mind what you have learned about weathering to help you answer the following questions.

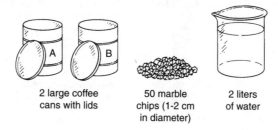

2 large coffee cans with lids

50 marble chips (1-2 cm in diameter)

2 liters of water

QUESTIONS

1. Which is the best prediction of the experiment's result?
   (1) The marble chips in can *A* will be smaller and rounder than the chips in can *B*.
   (2) The marble chips in can *B* will be smaller and rounder than the chips in can *A*.
   (3) There will be no difference between the marble chips in cans *A* and *B*.

2. Which graph best predicts what would happen to rocks in a fast-moving stream over time?

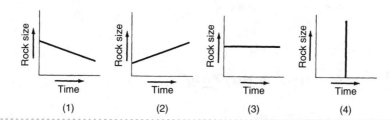

Review Questions

Multiple Choice

1. The lithosphere is the outer solid portion of the earth. It is most closely associated with the

    (1) crust   (2) mantle   (3) outer core   (4) inner core

2. An example of how the atmosphere interacts with the lithosphere occurs when

    (1) ocean water evaporates into the air

    (2) wind produces waves on the ocean

    (3) water evaporates from the ocean and goes into air

    (4) rain falls on a slope and washes soil away

3. When water freezes in cracks in a rock, the water expands, breaking the rock apart. This is a type of

    (1) glacial erosion   (3) chemical weathering

    (2) physical weathering   (4) groundwater erosion

4. The diagram below shows the mineral magnetite, which contains iron, changing into rust particles. This is an example of

Magnetite

Black, metallic, and magnetic

Black and rusty red, and less magnetic

Rusty red and nonmagnetic

    (1) physical weathering

    (2) chemical weathering

    (3) erosion by running water

    (4) the role of gravity in erosion

5. Erosion is the process by which rocks at Earth's surface

    (1) are removed and carried away

    (2) crumble and decay

    (3) turn into rust

    (4) melt to form magma

Thinking and Analyzing

6. How is weathering different from erosion?

7. What changed the shape of the land in the series of diagrams below?

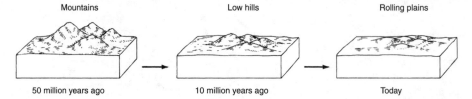

Mountains   Low hills   Rolling plains

50 million years ago   10 million years ago   Today

Internal Forces

Earth's internal forces also shape its surface. These forces produce *mountains*, *earthquakes*, and *volcanoes*, raising the land and building up Earth's surface.

Mountains, Plains, and Plateaus

1. *Mountains* are produced mainly by the processes of folding and faulting. **Folding** takes place when forces in Earth's crust press rocks together from the sides, bending the layers into folds. The land is squeezed into upfolds and downfolds, forming ridges and valleys (Figure 6-5).

    Faulting occurs when forces in the crust squeeze or pull rock beyond its capacity to bend or stretch. The rock then breaks and slides along a crack or fracture, called a *fault*, relieving the stress in the crust (Figure 6-6 on page 118). Faulting can produce mountains in a number of ways, as shown in Figure 6-7 on page 118.

2. Mountains can also be built by volcanoes. A **volcano** is a hole in Earth's crust through which lava flows from underground. During eruptions, the lava pours out onto the surface and cools to form solid rock, building upward in layers to produce a volcanic mountain, also called a volcano (see Figure 5-3 on page 85). Mount St. Helens in Washington state is a volcanic mountain.

    Besides mountains, other landforms that may result from uplift include plains and plateaus.

3. *Plains* are broad, flat regions found at low elevations. They are often made of layered sedimentary rocks that were formed underwater and slowly raised above sea level.

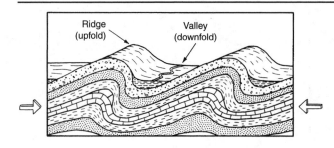

Ridge (upfold)   Valley (downfold)

Figure 6-5. *Folding: Forces in the crust can squeeze rock layers into folds.*

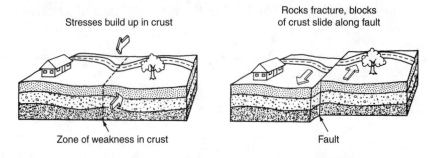

Stresses build up in crust

Zone of weakness in crust

Rocks fracture, blocks of crust slide along fault

Fault

Figure 6-6. *Faulting: When stresses in the crust reach the breaking point, the crust fractures and slips.*

4. *Plateaus* are large areas of horizontally layered rocks with higher elevations than plains. They can form in several ways. A large block of crust may rise up along faults to create a plateau, or a plateau may be gradually uplifted without faulting. Plateaus can also be built up by lava flows.

Earthquakes

Sudden movements of rocks sliding along faults in the crust are called *earthquakes*. Many earthquakes are associated with land uplift and mountain building. When an earthquake occurs it produces strong vibrations that travel through Earth.

Earthquakes produce three types of vibration waves. **Primary waves** (P-waves) and **secondary waves** (S-waves) travel through Earth, while **longitudinal waves** (L-waves) travel along Earth's surface. P-waves can travel through liquids and solids; however S-waves can only travel through solids. Analysis of how the three waves travel through Earth provides clues as to Earth's structure (Figure 6-8).

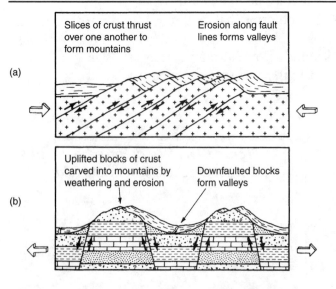

(a)

Slices of crust thrust over one another to form mountains

Erosion along fault lines forms valleys

(b)

Uplifted blocks of crust carved into mountains by weathering and erosion

Downfaulted blocks form valleys

Figure 6-7. *Mountains produced by (a) thrust faulting; (b) block faulting.*

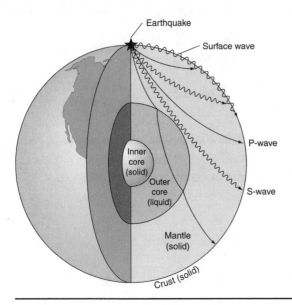

Figure 6-8. Paths of seismic waves traveling on the surface (surface waves) and through Earth (P- and S-waves)

Structure of Earth

The outer layer of Earth is called the **crust**. It is composed of solid rock material and forms a layer that covers Earth. Under the oceans the crust is about 5 kilometers thick and contains mostly basaltlike igneous rock. Under the continents the crust is about 50 kilometers thick and contains mostly granitelike igneous rock (Figure 6-9).

Below the crust is a layer called the **mantle**. It is about 2900 kilometers thick and probably consists of a dense iron- and magnesium-rich rock material. It is known to be solid because S-waves can travel through it, yet it flows very slowly and causes plates of the crust to move.

At Earth's center is the **core**, which is made up of an outer and inner zone. The *outer core* is about 2300 kilometers thick. It is thought to be liquid because S-waves cannot travel through it. The *inner core* has a radius of 1200 kilometers, and it is thought to be solid because P-waves travel faster through it. The core is believed to consist of mainly an iron and nickel mixture.

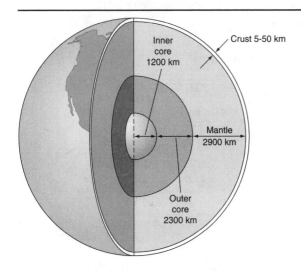

Figure 6-9. Cross section of Earth showing its internal structure.

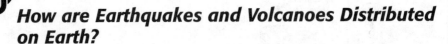

How are Earthquakes and Volcanoes Distributed on Earth?

Earthquakes and volcanoes are usually associated with the edges of crustal plates. At these locations the plates rub against each other (colliding or sliding) or up-welling (emerging) from under the crust. There are two major regions of earthquake and volcanic activity on Earth:

1) Circum-Pacific—around the Pacific Ocean, and
2) Trans-Mediterranean-Asia—across the Mediterranean Sea and Asia.

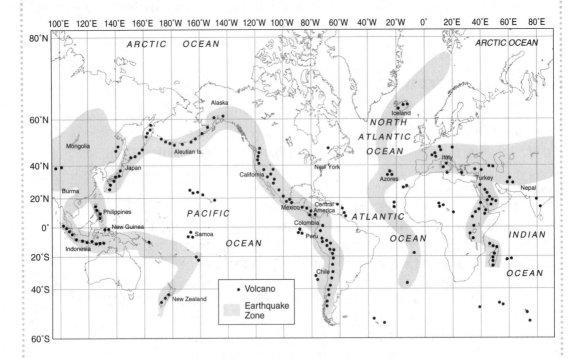

Using the world map above, locate the two zones.

Using the latitude and longitude of the volcanoes and earthquakes in the chart below, determine the location of each on the map.

Latitude and Longitude of 5 Volcanoes and 5 Earthquakes

Volcanoes	Lat.	Long.	Earthquakes	Lat.	Long.
1.	19 N	98 W	1.	31 S	99 W
2.	6 S	105 E	2.	34 N	139 E
3.	38 N	15 E	3.	53 N	168 W
4.	64 N	18 W	4.	36 N	36 W
5.	31 N	131 E	5.	11 N	85 W

(Continued)

1. Copy the chart below and fill in the name of the location for each earthquake and volcano you located on the map.

Volcano No.	Location	Earthquake No.	Location
1.		1.	
2.		2.	
3.		3.	
4.		4.	
5.		5.	

2. How many of the earthquakes and volcanoes are located in the Circum-Pacific or in the Trans-Mediterranean-Asia belt? How many are in neither belt?
3. Of the earthquakes or volcanoes that do not appear in either belt, do they appear to be associated with any other crustal plate boundaries? (See Figure 6-12 on page 125)
4. New York State is
   (1) in the Circum-Pacific belt and has many earthquakes and volcanoes
   (2) in the Trans-Mediterranean-Asia belt and has many earthquakes
   (3) not in either belt and has few earthquakes and no volcanoes
   (4) not in either belt and has few earthquakes and many volcanoes
5. Determine the location of five recent earthquakes and five recent volcanic events by gathering information from the Internet at the following URL addresses:

		Internet URL:
Earthquakes:	National Earthquake Information Center	<neic.usgs.gov>
Volcanoes:	Volcano World	<volcano.und.edu>

Determine the location of these earthquakes and volcanoes and whether they are in either belt.

Origin of Continental Drift

In 1912, Alfred Wegener, a German meteorologist, proposed that the continents were drifting across Earth's surface. He based his theory on how the shape of the continents fit together like pieces of a puzzle and on matching fossils, rocks, mountains, and glacial features

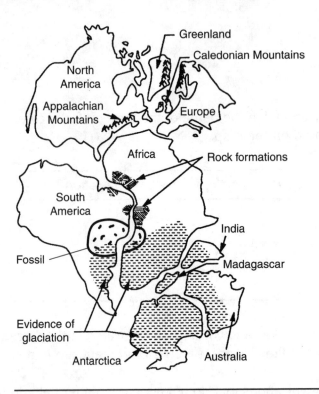

Figure 6-10. Continental shapes fit together like pieces of a puzzle. Fossils, rock formations, mountains, and glacial features match.

found on the continents on both sides of the Atlantic Ocean. (See Figure 6-10). He believed the continents were once connected in a single land mass that split apart and the pieces slowly drifted apart forming the Atlantic Ocean.

For about 50 years, the idea of drifting continents was rejected because no known mechanism or force within Earth could be identified as the cause of this massive movement in the crust. In the 1960s, the discovery of some ocean floor features supported the *continental drift* theory. Although the theory and evidence presented by Wegener was intriguing, only when the mechanism for movement was identified was the theory accepted.

Seafloor Spreading

In the 1960s, oceanographers discovered an underwater mountain ridge running north-south down the middle of the Atlantic Ocean. Along the ridge, there was much volcanic activity. It was suggested that new rock material was upwelling along the ridge. This rock moves east and west away from the ridge in a conveyor beltlike fashion, pushing out in opposite directions.

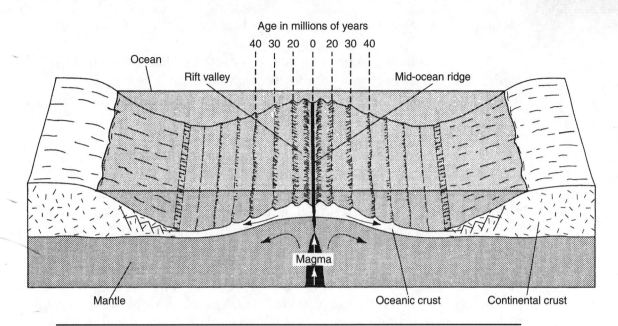

Figure 6-11. *Seafloor spreading: age of ocean crust increases with distance from mid-ocean ridge.*

In 1969, scientists discovered strips of ocean floor with matching magnetic polarity located on opposite sides of the Mid-Atlantic Ridge. This provided evidence that *seafloor spreading* was occurring (see Figure 6-11). Other evidence in the form of matching the type and age of sediments and fossils on each side of the ridge, determining the ridge as a high heat flow area, and trenches at locations where crust was subsiding also helped to support the concept of seafloor spreading.

The internal mechanism causing the continents to drift and seafloor to spread appears to be heat flowing upward from inside Earth. The upwelling carries new rock to the surface, slides the continents apart, and in some locations the crust eventually subsides and forms a trench in the ocean floor.

Plate Tectonics

There is much evidence that forces at work inside Earth have raised the level of the land. For example, many mountaintops are made of sedimentary rock that was formed originally on the ocean floor. Folds and faults seen in many rock outcrops are also signs of crustal movements caused by internal forces. Scientists explain these forces and the movements they produce by the theory of *plate tectonics*.

According to this theory, Earth's crust is broken up into a number of large pieces, or *plates*, that slowly move and interact in various ways. Some plates are spreading apart, some are sliding past each other, and some are colliding. These movements cause mountain building, volcanic activity, and earthquakes along the plate edges. Figure 6-12 shows Earth's major crustal plates.

Scientists believe that plate motions are caused by heat circulating in Earth's *mantle*, the thick zone of rock beneath the crust. The heat softens mantle rock so that it flows very slowly, following the heat currents and carrying along overlying pieces of crust (see Figure 6-13).

The processes of plate tectonics create many of Earth's surface features. The collision of two plates carrying continents produces great mountain ranges. The Himalayas were formed in this way.

When one plate slides sideways past another plate, a major fault and earthquake zone is produced. In California, the Pacific Plate is sliding past the North American Plate along the San Andreas Fault, sometimes causing severe earthquakes.

Where plates are spreading apart, ocean basins are formed. Large continents are broken into smaller land masses that move away from each other in a process called *continental drift*. This is taking place today where the Arabian Plate is splitting away from the African Plate, opening up the Red Sea.

Ocean Floor Features

Almost three-quarters of Earth's surface is covered by ocean water. The floor of the ocean is not all flat and featureless. Scientists have found that the ocean floor has mountains, valleys, plains, and plateaus. Many of these features, such as mid-ocean ridges and ocean trenches, are produced by the processes of plate tectonics.

1. A *mid-ocean ridge* is a long, underwater mountain chain where rising magma forms new ocean crust. The new crust is added to crustal plates that spread away from the ridge, as shown in Figure 6-13. This process is called *seafloor spreading*.

2. *Trenches* are underwater valleys that form the deepest parts of the ocean floor. A trench is found where a plate of ocean crust collides with another plate and is forced to slide under it, back into Earth's mantle. This causes volcanic activity and mountain building along the edge of the upper plate (see Figure 6-13).

   Other ocean floor features include continental shelves, continental slopes, the deep ocean floor, and seamounts.

3. *Continental shelves* are areas of the seafloor that slope gently away from the coastlines of most continents. The angle of slope is so slight that if you could stand on a continental shelf, you would think you were on level ground.

4. *Continental slopes* drop away from the outer edges of continental shelves to the great depths of the ocean. These slopes are much steeper than continental shelves.

   Continental slopes level off into the *deep ocean floor*. The deep ocean floor is not simply a flat plain; it also has ridges and valleys.

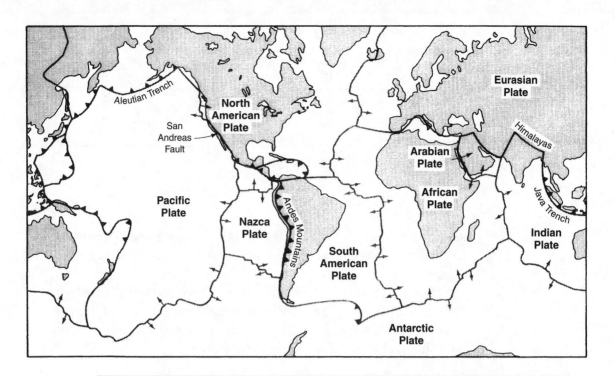

Figure 6-12. *Earth's major crustal plates. (Arrows show where plates are spreading apart, triangular "teeth" show where one plate is sliding beneath another plate.)*

Figure 6-13. *Plate tectonics: Heat currents in the mantle cause movements of Earth's crustal plates, producing many features on the seafloor and the continents.*

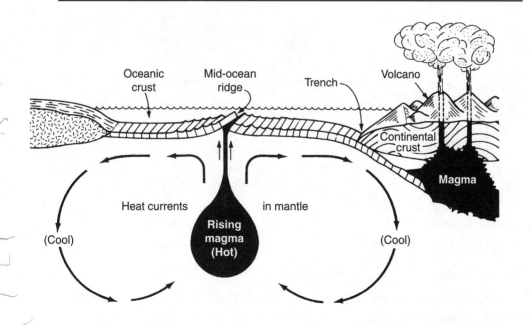

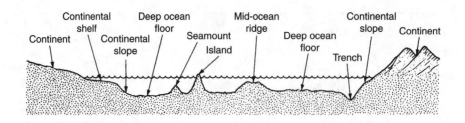

Figure 6-14. *Features of the ocean floor.*

Rising here and there from the ocean floor are tall underwater mountains called *seamounts*. Most seamounts were formed by volcanoes.

When the top of a seamount rises above the water's surface, an island is formed. The Hawaiian Islands are the tops of a chain of volcanic seamounts. Figure 6-14 shows the profile of an ocean floor that includes many of these features.

 Review Questions

Multiple Choice

8. The rock layers in the diagram have been affected by

    (1) volcanoes
    (2) faulting
    (3) groundwater erosion
    (4) folding

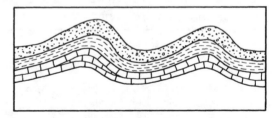

9. Mountains can be produced by all of the following processes *except*

    (1) volcanic eruptions   (3) weathering
    (2) folding   (4) faulting

10. The theory that Earth's crust is broken up into large pieces that move and interact is called

    (1) evolution   (3) the rock cycle
    (2) mountain building   (4) plate tectonics

11. Major mountain ranges are formed when crustal plates

    (1) push into each other   (3) move away from each other
    (2) slide past each other   (4) break into smaller plates

12. If crustal block *A*, to the left of the fault in the diagram, suddenly shifted downward several feet what would most likely occur at location *C*?

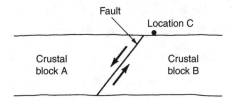

(1) An earthquake would occur.

(2) A volcanic eruption would occur.

(3) A mountain would form.

(4) An ocean would form.

13. The ocean floor is best described as

(1) a flat, featureless plain

(2) having mountains, valleys, plains, and plateaus

(3) a flat plain with a deep valley in the center

(4) having plains and plateaus only

14. The table at the right shows the depth of the ocean at various distances from a continent. At what distance from the continent is a deep trench located?

(1) 100 kilometers

(2) 200 kilometers

(3) 300 kilometers

(4) 500 kilometers

Distance from continent	Ocean depth
50 km	400 m
100 km	9000 m
150 km	1250 m
200 km	1100 m
250 km	200 m
300 km	950 m

15. Earth's crust is in motion. The Mid-Atlantic Ridge is a location of up-welling magma where new crust is being formed and pushed outward in two directions, east and west. Which of the following locations would have the youngest rocks?

(1) A

(2) B

(3) C

(4) F

16. In which state is an earthquake most likely to occur?

(1) New York   (2) Florida   (3) Kansas   (4) California

17. Alaska is more likely to have an earthquake than Florida because

   (1) Alaska is younger than Florida

   (2) Alaska is in the middle of a crustal plate

   (3) Alaska is on the edge of a crustal plate

   (4) Florida has more people

18. Earth is made up of a crust, mantle, outer core, and inner core. The crust is solid and broken into plates that seem to float on the mantle. The mantle is solid, but flows like heated plastic. The outer core is liquid, and the inner core is solid. Much of our knowledge about the internal structure and composition of Earth comes from

   (1) deep mining operations   (3) underwater drilling

   (2) earthquake produced waves   (4) the rock structure on the surface

19. Early recognition of Continental Drift came from

   (1) the land on each side of the Atlantic Ocean fitting like a puzzle

   (2) identifying earthquake and volcanic belts

   (3) making measurements of the distance across the Atlantic Ocean

   (4) determining the age of rocks under the Atlantic Ocean

Thinking and Analyzing

20. Earthquakes can occur anywhere on Earth's surface; however they occur much more frequently on the edge of crustal plates. A very high percentage of earthquakes occur in the Circum-Pacific belt and the Trans-Mediterranean-Asia belt. Mid-ocean ridges are also locations where earthquakes are more likely to happen.

   Copy the tables below and record your answers in those tables. Using the latitude and longitude of the two earthquakes listed below, locate the position of each on the accompanying world map. Also, identify which earthquake belt each is associated with.

Latitude	Longitude	Locations	Earthquake Belt
36° N	31° E		
34° N	140° E		

Find earthquake locations A and B on the map. What is the latitude and longitude of these two earthquakes? With what crustal boundary are they each associated?

Position	Latitude	Longitude	Crustal Boundary
A			
B			

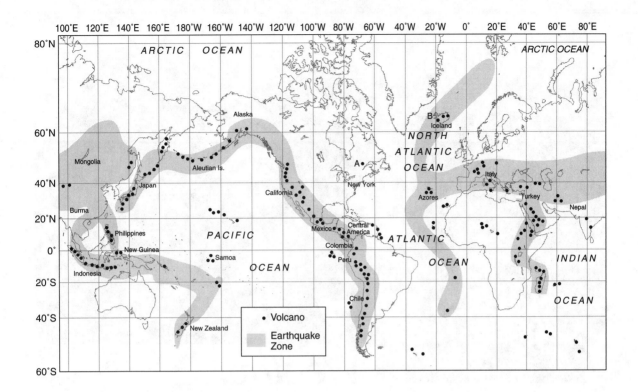

Use the map below to assist you with questions 21 to 24.

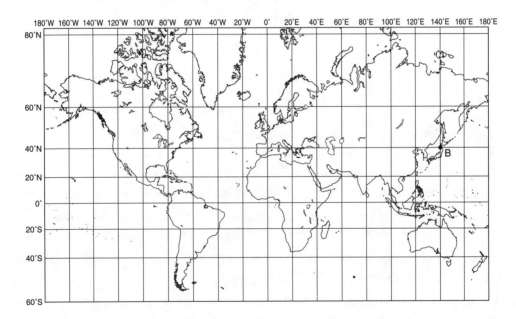

21. An earthquake occurred at 0° latitude and 20° W longitude. With what ocean floor crustal feature was this earthquake associated?

22. Describe what processes most likely caused this earthquake to occur.

23. A volcanic eruption occurred at Point B on the map. What is the latitude and longitude of the eruption?

24. With what earthquake belt is this volcanic eruption associated?

Chapter 7
Weather and Climate

Points to Remember

> Weather is the present state of the atmosphere at a given location for a short period of time. Climate is the average weather that prevails over a large area for a long period of time.

> The atmosphere consists of mostly nitrogen and oxygen. It can be divided into layers by temperature. Almost all weather takes place in the lowest layer, the troposphere.

> The sun is the primary source of energy affecting Earth's surface. Weather changes occur due to the uneven heating of Earth's surface by the sun.

> Air masses are large bodies of air with similar temperature and humidity conditions. Most weather changes are due to the movement of air masses.

> Air pressure systems are either high- or low-pressure systems. A high-pressure system brings fair weather; a low-pressure system brings stormy weather.

> Weather maps show the positions of air masses, fronts, and weather elements at many locations.

130

Defining Weather

Surrounding Earth is a layer of gases called the atmosphere. These gases make up what is commonly known as *air*. **Weather** consists of the conditions of the atmosphere, such as temperature, humidity, precipitation, wind, and cloud cover. These conditions change from day to day and from place to place. Energy from the sun is the main cause of these changes.

Structure of the Atmosphere

The atmosphere consists of gases that are held to Earth's surface by gravity. About 97 percent of the gases are located in the lowest 30 kilometers above the surface. Although there is no boundary to the upper limit of the atmosphere, gases thin sufficiently to suggest the top of the atmosphere to be about 150 km above the surface.

   Scientists use the temperature changes that occur with increasing altitude as a means of identifying four layers to the structure of the atmosphere. (See Figure 7-1) The lowest layer, the troposphere, is about 10 kilometers thick. Nearly all weather takes place in the *troposphere*, and it supports life

Figure 7-1. *Layers of Earth's atmosphere as described by temperature changes.*

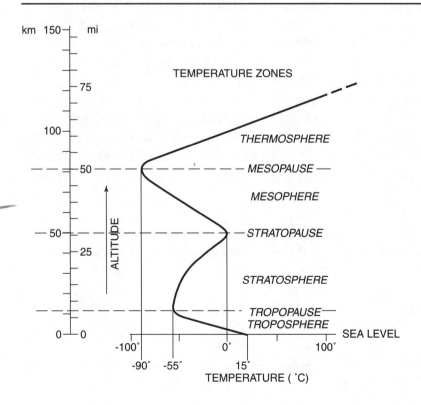

with sufficient amounts of oxygen, water vapor, and carbon dioxide. An increase in altitude reveals a decrease in temperature. At the top of the troposphere there is a thin layer with a constant temperature of about $-55°C$ called the *tropopause*.

The second layer, the *stratosphere* extends to an altitude of 50 kilometers and increases in temperature with an increase in altitude. The upper boundary of the stratosphere is the *stratopause*, which has a constant temperature of about $0°C$. The third layer is the *mesosphere* that decreases in temperature with an increase in altitude. Above the mesosphere is the *mesopause*, a layer with a constant temperature of about $-90°C$. The uppermost layer is the *thermosphere* which increases in temperature upward.

Composition

Table 7-1 provides the average composition of Earth's atmosphere. The composition of the atmosphere is relatively constant. The interactions of the lithosphere, hydrosphere, and atmosphere assist in replenishing gases through cycles such as the water cycle, the nitrogen cycle, and the oxygen-carbon dioxide cycle.

Weather Elements

Weather is made up of a number of elements, including *air temperature*, *air pressure*, *humidity*, *wind speed* and *direction*, *clouds* and *cloudiness*, and *precipitation*. The weather at any given location can be described in terms of its elements.

1. *Air temperature* indicates the amount of heat in the atmosphere. It is measured with a *thermometer*. The two temperature scales commonly used for thermometers are the *Celsius* scale and the *Fahrenheit* scale, shown and compared in Figure 7-2. Both scales are divided into units called *degrees*.
2. Because air has weight, air presses down on Earth's surface with a force called **air pressure**. This force is measured with a *barometer*, in either inches of mercury or millibars. Variations in air pressure are caused mainly by temperature and altitude. Temperature affects air pressure because a given volume of cool air weighs more than

Table 7-1. Composition of the Atmosphere

Nitrogen	78%
Oxygen	21%
Carbon dioxide	0.03%
Other gases	0.17%
Water vapor (variable)	1–3%

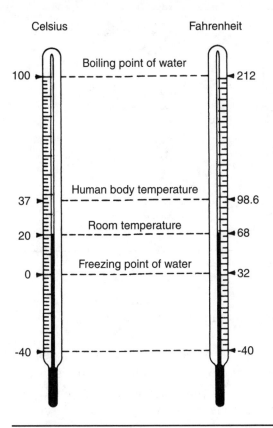

Figure 7-2. *Temperature scales compared.*

an equal volume of warm air. Altitude affects air pressure because places at high elevations, being farther up in the atmosphere, have less air pressing down on them. Therefore, as altitude increases, air pressure decreases.

3. **Humidity** is the amount of *water vapor* (water in the form of a gas) present in the air. Warm air can hold more water vapor than cool air can. *Relative humidity* is the ratio between the actual amount of water vapor in the air and the maximum amount of water vapor the air can hold at that temperature. For example, a relative humidity of 90 percent means that the air contains 90 percent of the water vapor it can hold at its current temperature. Relative humidity is measured with a *hygrometer* (sometimes called a wet-and-dry-bulb thermometer).

4. **Wind** is the movement of air over Earth's surface. *Wind speed* is a measure of how fast the air is moving, in miles or kilometers per hour. It is determined with an *anemometer*. **Wind direction** is the direction *from which the wind is coming*; it is NOT the direction the wind is blowing *toward*. For instance, a wind blowing from north to south is called a north wind. Wind direction is determined with a *wind vane*, which points in the direction from which the wind is blowing.

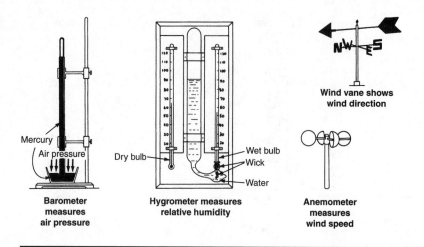

Figure 7-3. Weather instruments.

Figure 7-3 shows some instruments used to measure weather elements.

5. **Clouds** are masses of tiny water droplets or ice crystals, suspended in the air. *Cloudiness,* the amount of sky covered by clouds, is observed directly and described with phrases like "partly cloudy" or "mostly cloudy." A sky completely covered by clouds is described as "*overcast.*"

6. **Precipitation** is water, in any form, falling from clouds in the sky. Precipitation can fall as rain, snow, sleet, or hail. A *rain gauge* is used to measure the amount of precipitation in inches.

 *Laboratory Skill*

How Do You Read Weather Instrument Scales?

Instruments are used to measure weather elements such as temperature and air pressure. Each of these two elements has more than one unit of measurement. A description of the units used for each instrument follows:

Thermometer: measures air temperature in degrees Celsius or degrees Fahrenheit. If the scale is calibrated for Celsius, the number scale should start at about −40°C and go to 50°C, and if it is Fahrenheit, the number scale should start at about −40°F and go to 120° F. Although weather data are usually given in degrees Fahrenheit, Celsius readings are slowly replacing the Fahrenheit units.

Barometer: measures air pressure commonly in either inches of mercury or millibars. If the scale is in inches of mercury, weather readings generally fall between 29.00 and 31.00 inches of mercury. When the scale is in millibars, readings fall between 980 and 1050 millibars. Inches of mercury are still used in weather forecasting; however, weather maps use millibars.

(Continued)

The nonliquid thermometer or barometer has a dial with a scale that indicates the present reading.

The thermometer scale below has a range that goes from −30 to 50. Between each numbered line there are 10 degrees. Half way between 20 and 30 is 25 degrees. Can you find 27 degrees?

The barometer scale below has a range that goes from 29.00 to 31.00. Each numbered line indicates 0.10 inch; therefore the first numbered line after 29.00 is read 29.10 inches of mercury. The lines that are not numbered indicate 0.05 inch. The next line after 29.10 is read as 29.15 inches of mercury. Can you find 29.17 inches of mercury?

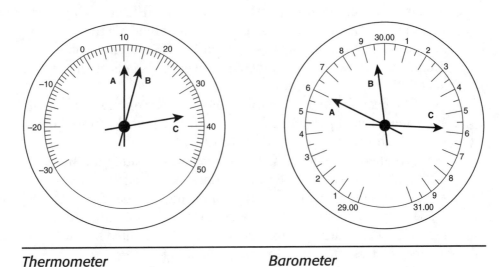

Thermometer   Barometer

QUESTIONS

1. The thermometer scale above indicates the units of measurement to be
   (1) degrees Celsius   (3) millibars
   (2) degrees Fahreinheit   (4) inches of mercury
2. The barometer scale above indicates the units of measurement to be
   (1) degrees   (2) miles per hour   (3) millibars   (4) inches of mercury
3. Copy the table; record the measurements indicated by pointers A, B, and C.

Instrument	Pointer A	Pointer B	Pointer C
Thermometer			
Barometer			

4. Copy the thermometer above and locate the following readings on the scale.
   (1) 5°C   (2) −15°C   (3) 22°C
5. Copy the barometer above and locate the following readings on the scale.
   (1) 29.20 inches   (2) 30.15 inches   (3) 30.42 inches

The Sun's Energy

The sun is the main source of the energy in the atmosphere. As the sun heats Earth's surface, the surface radiates this heat energy back into the atmosphere. However, the sun does not heat Earth's surface evenly; consequently, the atmosphere is not heated evenly. This uneven distribution of heat energy in the atmosphere is the cause of weather.

The heating of Earth's surface depends to some extent on the nature of the surface, since some kinds of surfaces get hotter than others. For instance, pavement and sand get much hotter than do grass and water. On a larger scale, the surfaces of oceans, forests, and deserts are also affected differently by the sun. These surfaces, in turn, heat the air above them differently, producing variations in air temperature.

When air is heated, it becomes lighter (less dense) than the surrounding air. Therefore, warm air rises. Cool air is heavier (more dense), so it tends to sink. As air rises or falls, the surrounding air rushes in to replace it, causing air to circulate, as shown in Figure 7-4. This circulation, which can take place over a few kilometers or over thousands of kilometers, brings about changes in the weather.

The heating of Earth's surface also depends on the angle at which the sun's rays strike the surface (Figure 7-5). Near the equator, the sun's rays strike Earth vertically or nearly vertically (Area *A* in Figure 7-5). This concentrates the sun's energy within a small area, heating the surface very effectively. However, since Earth's surface is curved, the sun's rays strike areas away from the equator at a slanting angle (Areas *B* and *C* in Figure 7-5). This spreads energy over a wider area, heating Earth's surface less effectively. The farther from the equator, the more slanted the sun's rays come in, and the less effectively they heat the surface.

The uneven heating of Earth's curved surface causes hotter air at the equator to rise and spread to the north and south, while cooler air near Earth's poles moves toward the equator to replace the rising air. Earth's

Figure 7-4. *Air circulation pattern.*

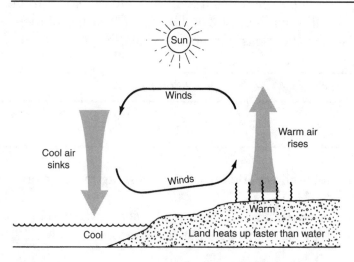

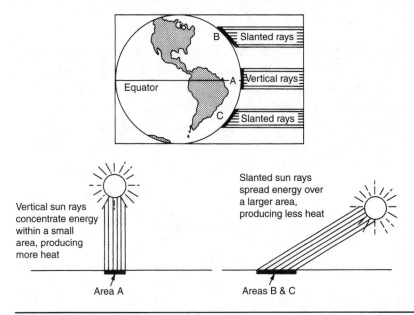

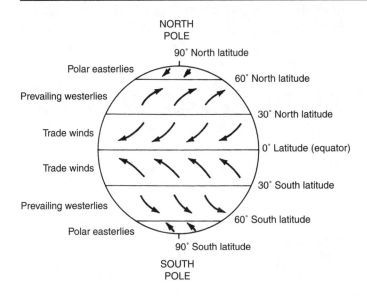

Figure 7-5. *The angle of the sun's rays affects the heating of Earth's surface.*

rotation breaks up this simple circulation into complex global *wind belts*, shown in Figure 7-6, in which winds blow in different directions at different *latitudes*.

Latitude is distance north or south of the equator. The lines drawn parallel to the equator on a globe or a map are lines of latitude. Winds that commonly blow in the same direction at a given latitude are called **prevailing winds**. Most of the United States lies in the Northern Hemisphere between 30° and 60°N latitude within the *prevailing westerlies*. These winds cause most weather systems to move from west to east.

Figure 7-6. *Earth's wind belts.*

The Water Cycle

The sun's energy also powers the **water cycle**. In this process, water circulates through the atmosphere, lithosphere, and hydrosphere, as shown in Figure 7-7. Heat from the sun changes liquid water into water vapor. This is called **evaporation**. For example, evaporation takes place when a puddle of rainwater shrinks and dries up on a hot, sunny day. Water enters the atmosphere by evaporation from oceans, lakes, and rivers, and by plants releasing water vapor through their leaves.

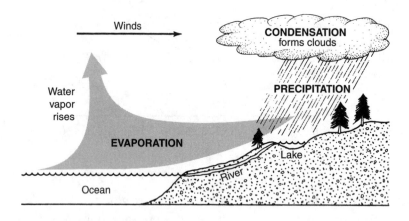

Figure 7-7. *The water cycle.*

Warm air can hold more water than cool air can. Therefore, when air cools, it cannot hold as much water vapor, and some of it changes back into droplets of liquid water. This process, called **condensation**, produces dew, fog, and clouds.

1. *Dew* is formed when water vapor condenses onto cool surfaces or objects. This takes place, for instance, when water droplets appear on the outside of a glass that holds a cold drink (see Figure 7-8)
2. Condensation may also form tiny water droplets that remain suspended in the air. When this takes place near the ground, *fog* is produced.

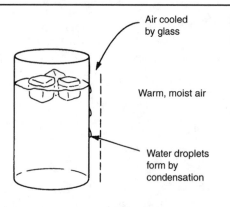

Figure 7-8. *Condensation occurs on a cold glass.*

3. When moist air rises high into the atmosphere and cools, condensation forms *clouds*. If enough water vapor condenses, the tiny water droplets may join together into larger, heavier drops that fall to Earth as precipitation. Then the cycle of evaporation, condensation, and precipitation can begin all over again.

Climate

Climate is the general character of the weather that prevails in an area from season to season and from year to year. It can be thought of as the average weather of an area over a long period of time.

A number of factors combine to produce different climates.

1. One factor is **latitude**, which is distance from the equator. Places at high latitudes, far from the equator, tend to have colder climates than places at lower latitudes. For instance, Canada is at a higher latitude than Mexico, so it has a colder climate.
2. Another factor is **altitude**, which is the height (elevation) above sea level of a place. Higher elevations are cooler than lower elevations. Just as a mountaintop is colder than its base, a city at a high altitude will have a colder climate than a city at the same latitude that is at a lower altitude.
3. Large bodies of water can affect climate. Land areas close to oceans or large lakes generally have more moderate climates (cooler summers and warmer winters) than areas far from water. Water absorbs and gives off heat more slowly than land does. Therefore, as the land heats up during summer, the water stays relatively cool, keeping coastal areas cooler in summer than places farther inland. In winter, the situation is the opposite. The water loses heat built up during summer more slowly than the land does, keeping coastal areas warmer in winter than areas farther inland.
4. Mountain barriers can also influence climate. The side of a mountain range facing the prevailing winds tends to have a cool, moist climate, while the opposite side of the mountains has a warmer, drier climate. This is illustrated in Figure 7-9.

Figure 7-9. *A mountain range can affect climate.*

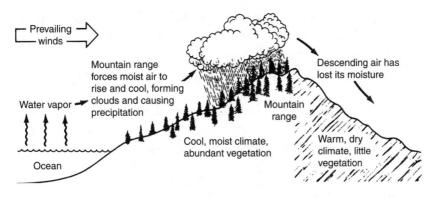

Designing an Experiment; Predicting Results

Cold air is heavier and more dense than warm air. Therefore, cold air tends to sink and warm air tends to rise. To design an experiment that can demonstrate this, you would need cool and warm air, a way to control the flow of air, and several thermometers.

For instance, a small room with a window provides a means of controlling air flow (see the diagram). On a day when the temperature outside is lower than the indoor temperature by at least 10°C, you could open the window to let in cold air and measure temperature changes in the room with thermometers. If it is true that cold air sinks and warm air rises, the lower levels of the room will get colder more quickly than the higher levels will. Answer the questions that follow the diagram on this page.

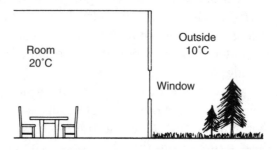

QUESTIONS

1. How should the thermometers be arranged in the room to show that cold air sinks and warm air rises?

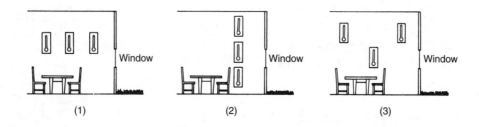

2. If this experiment were performed on a summer day, when the outdoor temperature was 10°C higher than the indoor temperature, what would probably happen in the room after opening the window?
(1) The temperature near the ceiling would increase faster than the temperature near the floor.
(2) The temperature near the floor would increase faster than the temperature near the ceiling.
(3) The temperature at all levels of the room would increase at the same rate.

Multiple Choice

1. What temperature is indicated by the thermometer in the diagram?

   (1) 10°C

   (2) 20°C

   (3) 25°C

   (4) 30°C

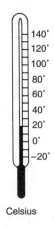

Celsius

2. In the diagram, the most likely reason for City *B* to be cooler than City *A* is that

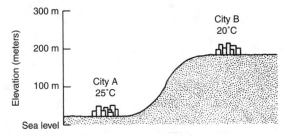

   (1) City *B* is at a different latitude   (3) City *B* is closer to the ocean

   (2) City *B* is at a higher altitude   (4) City *B* is closer to the sun

3. The diagram below shows two cities, *A* and *B*, and their positions on a continent. How will the climates of the cities compare?

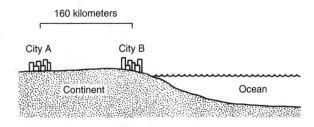

   (1) City *B* will have warmer summers and colder winters than City *A*.

   (2) City *B* will have cooler summers and warmer winters than City *A*.

   (3) City *B* will have cooler summers and colder winters than City *A*.

   (4) Both cities will have the same climate.

4. The most likely reason that New York City has a cooler climate than Miami is

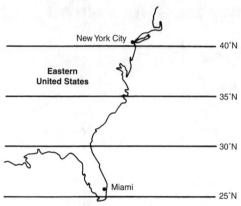

(1) a difference in distance from the ocean

(2) a difference in altitude

(3) a difference in air pressure

(4) a difference in latitude

5. Which graph indicates the general temperature change as you travel from the North Pole (NP) to the South Pole (SP)?

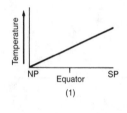

6. Which letter in the water-cycle diagram indicates where condensation occurs?

(1) *A*

(2) *B*

(3) *C*

(4) *D*

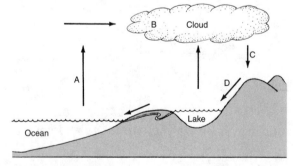

7. The source of the energy that sets Earth's atmosphere in motion and causes weather is

(1) volcanism (2) gravity (3) the ocean (4) the sun

Questions 8–11 refer to the data and diagram below, which represents a beach on a hot summer day at about 2:00 P.M.

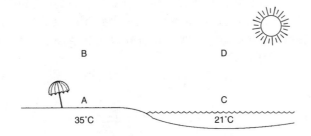

A. the air one meter above the beach sand

B. the air at 1500 meters above the beach sand

C. the air one meter above the ocean water

D. the air at 1500 meters above the ocean water

8. Rising warm air is moving between locations

   (1) A & B (2) C & D (3) A & D (4) C & B

9. There is sinking cool air moving between locations

   (1) B & A (2) D & C (3) B & C (4) D & A

10. There is horizontally moving cool air between locations

    (1) B & D (2) A & C (3) D & B (4) C & A

11. Clouds form from rising air. Clouds may be forming at location

    (1) A (2) B (3) C (4) D

12. The two temperature scales commonly used for thermometers are the Celsius scale (°C) and the Fahrenheit scale (°F), shown and compared in the following diagram. Both scales are divided into units called degrees. The two thermometers are aligned so that you can easily convert one scale to the other scale by using a ruler or the edge of a sheet of paper.

    Using the thermometer diagram, convert each of the following temperature readings:

    104°F = ___°C   50°C = ___°F

    −40°F = ___°C   75°C = ___°F

    75°F = ___°C   20°C = ___°F

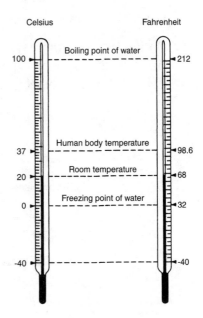

Thinking and Analyzing

13. Mr. Chang decided to drive his car up Pikes Peak, a tall mountain in Colorado. He has a thermometer and a barometer in the car. Describe how the temperature and air pressure readings would change as he drove up the mountain.

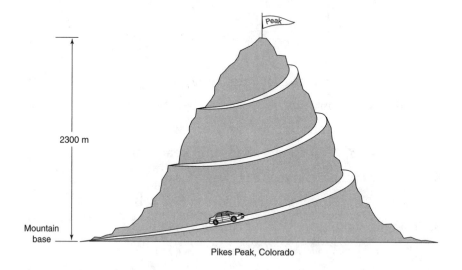

Pikes Peak, Colorado

14. After a rain shower on a summer day the sun came out and the puddle in front of Aldo's house evaporated quickly. A few days later another rain shower occurred and the sky remained cloudy. The puddle took much longer to evaporate. Aldo believes the sun caused the puddle to evaporate quickly. Design an experiment to test Aldo's hypothesis. Include the following information:

Clearly state the hypothesis.

State the factor to be changed.

State two constant factors that will not be changed.

Describe the procedure.

15. On a sunny summer day, a thermometer was placed above each of the surfaces shown in the diagram. Describe the differences expected in each of the thermometer readings.

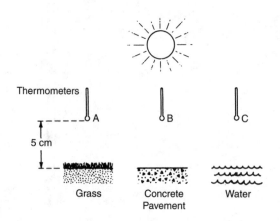

16. Briefly explain why some people might spend their summer vacation near an ocean or a lake.

17. Describe the difference between weather and climate.

Large-Scale Weather Systems

Air Masses

A large body of air that has uniform temperature and moisture conditions throughout it is called an **air mass**. Much of our weather is determined by air masses.

An air mass forms when air stays over a large area of Earth's surface and takes on the temperature and moisture characteristics of that area. An air mass that forms over a warm body of water, like the Gulf of Mexico, will be warm and moist. Air masses that enter the United States from Canada are usually cold and dry because they formed over a cool land surface. An air mass builds up over an area for a few days and then begins to drift across Earth's surface.

There are four different surface conditions that affect the formation of air masses. Air masses that originate over *land* are dry. Those that form over *water* are moist. Cold air masses originate near the *poles*, and warm air masses form near the *equator*.

The major air masses that affect the continental United States, shown in Figure 7-10, enter the country from the north, west, and south. They are then blown from west to east by the prevailing winds. As an air mass

Figure 7-10. *Major air masses affecting the United States.*

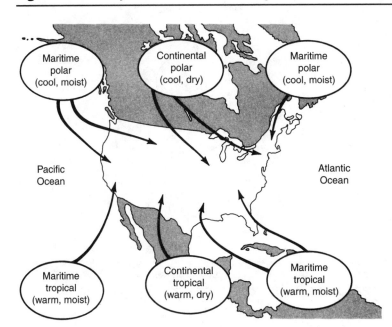

moves, it changes the local weather conditions at the surface below. The weather may become warmer or cooler, wetter or drier, depending on the type of air mass passing by.

High- and Low-Pressure Systems

Surface air pressures are usually highest in the centers of air masses, so these areas are called *high-pressure systems*, or simply *highs*. The air in a high tends to sink, and winds blow outward from the center, turning in a clockwise direction, as in the Northern Hemisphere, shown in Figure 7-11(a). High-pressure systems usually bring clear skies, dry weather, and gentle winds.

Surface air pressures are lower toward the edges of air masses. These areas form *low-pressure systems*, or *lows*. The air in a low tends to rise, and winds spiral in toward the center in a counterclockwise direction in the Northern Hemisphere, as shown in Figure 7-11(b). (Note that in both highs and lows, winds always blow from areas of higher air pressure toward areas of lower air pressure.) Low-pressure systems usually bring cloudy, wet weather, often with strong, gusty winds.

Highs and lows are generally indicated on weather maps by the letters **H** and **L** (see Figure 7-14 on page 149). The highs on weather maps usually indicate the centers of air masses. A low forms between highs, much like a valley between two mountains. Highs are large and tend to move slowly, therefore, weather changes are usually gradual, rather than sudden. Changes in air pressure readings signal the passing of highs and lows.

Figure 7-11. *Movement of air currents in (a) a high-pressure system; (b) a low-pressure system (mb stands for millibars).*

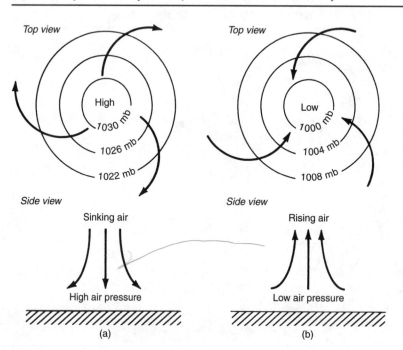

Organizing Information into a Table

Information can often be put into an orderly format by creating a chart or table that summarizes it. For instance, information about air masses can be organized in the form of a table. Air masses are named for the temperature and moisture characteristics of the region over which they form. *Continental* air masses form over land and are dry. *Maritime* air masses form over water and are moist. *Tropical* air masses form in the south and are warm, while *polar* air masses form in the north and are cool. For example, an air mass that formed over Canada would be called *continental polar*, indicating that it is cool and dry.

Types of Air Masses

Name of Air Mass	Temperature (Warm or Cool)	Humidity (Moist or Dry)
Continental Polar	Cool	Dry
Maritime Polar	?	Moist
?	Warm	Dry
Maritime Tropical	?	?

The chart above represents some of this information. Copy the chart and fill in the missing information.

Based on your completed table, how would you describe a *continental tropical* air mass in terms of its temperature and moisture content?

The following incomplete table contains information about air pressure systems. Copy the table and refer to the section in the text about highs and lows to help you fill in the missing information. Then answer the questions below.

Air Pressure Systems

Type of Air Pressure System	Characteristics			
	Vertical Air Movement	Horizontal Air Movement	Type of Weather	Location in Air Mass
High	Sinking	?	Fair	?
Low	?	In toward center, turning counter-clockwise	?	Edges

QUESTIONS

1. The movement of air in a high-pressure system is
   - (1) counterclockwise and rising
   - (2) counterclockwise and sinking
   - (3) clockwise and rising
   - (4) clockwise and sinking
2. The weather on the edge of an air mass is usually
   - (1) stormy, with low air pressure
   - (2) stormy, with high air pressure
   - (3) fair, with low air pressure
   - (4) fair, with high air pressure

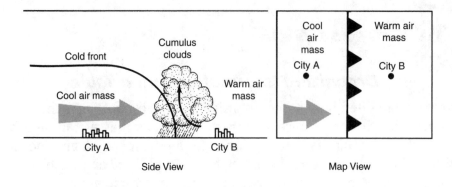

Figure 7-12. Cold front.

Fronts

When one air mass comes into contact with another air mass, a boundary, called a *front*, forms between them. Sudden changes in weather conditions can occur from one side of a front to the other. The air masses that meet often differ in temperature, humidity, and density. These differences prevent the air masses from mixing. The cooler, drier air is heavier and remains close to the ground, while the warmer, moister air is lighter and rises upward. This causes areas of low pressure to develop along fronts, often producing clouds, strong winds, and precipitation. These lows are the major storm systems of our latitudes.

Different kinds of fronts are produced depending on how the air masses come together.

1. *Cold Front.* If a cold air mass pushes into and under a warm air mass, a **cold front** is formed (Figure 7-12). Cold fronts usually bring brief, heavy downpours, gusty winds, and cooler temperatures. On hot, humid summer days, the passing of a cold front typically causes thunderstorms, followed by a decrease in temperature and humidity.
2. *Warm Front.* When a warm air mass pushes into and over a cold air mass, a **warm front** is created. The warm air slides up and over the cooler air (Figure 7-13). Warm fronts bring light precipitation lasting

Figure 7-13. Warm front.

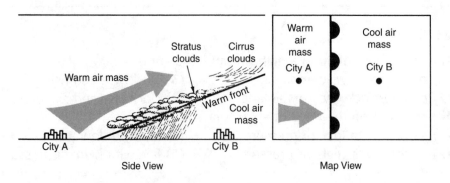

a day or two and warmer temperatures. When the sky is overcast all day with light rain falling, a warm front is most likely present.

Different types of clouds are produced depending upon how air rises. If air is pushed straight up, as it is along a cold front, puffy *cumulus* clouds are produced. If air rises at a low angle, as it does along a warm front, flat layers of *stratus* clouds are formed. Wispy *cirrus* clouds, which are made up of ice crystals, form high in the atmosphere. They may look like feathers, or tufts of hair. The presence of many high cirrus clouds may indicate that a warm front is approaching (see Figure 7-13).

Weather Forecasting

Weather forecasting is an attempt to make accurate predictions of future weather. The accuracy of weather forecasting is improving as technology advances. In addition to weather balloons, thermometers, and barometers, weather forecasters now have a wide array of weather satellites, radar devices, and computer systems at their disposal.

Short-range local forecasts are comparatively easy. They are based mostly on *air pressure* readings and observations of *cloudiness* and *wind direction*. Changes in these weather elements are usually good indications of the weather for the next day or two.

(1) Decreasing air pressure readings signal the approach of stormy weather, while rising air pressure suggests that fair weather is coming. (2) An increase in cloudiness is a sign that a front is approaching, probably bringing precipitation. (3) In New York State, winds blowing from the west usually bring fair weather, while winds from the south or east often bring wet weather.

Today, weather forecasters also use information from weather satellites and radar to improve their short-range forecasts. This information is used to produce up-to-date weather maps like the one in Figure 7-14. Such maps can

Figure 7-14. *A weather map is useful for making predictions.*

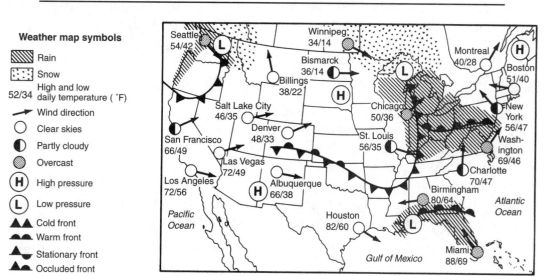

help us predict coming changes. Weather systems generally move from west to east across the United States. Therefore, if a weather map shows a high immediately to our west, we can forecast fair weather for the next day. Conversely, if the map shows a low to our west, we can expect stormy weather.

Long-range weather forecasting is more difficult. Computers, satellite photographs, and radar images enable forecasters to track the movements of large-scale weather systems like air masses and fronts. With this information, they can make predictions of weather several days in the future. However, such forecasts are not always accurate. Generally, weather forecasts for the next day or two are reliable, while long-range forecasts usually have to be revised.

Process Skill

Weather Forecasting

The science club wants to start a weather station for the purpose of forecasting local weather for after-school events. The chart shows what information needs to be collected, what instruments are needed, and the units of measurement for each instrument.

Information	Instrument	Measurement Units
Temperature	Thermometer	Degrees Celsius
Air Pressure & Tendency	Recording Barometer	Millibars
Humidity	Hygrometer	Percent
Wind Speed	Anemometer	Miles Per Hour
Wind Direction	Wind Vane	Compass Direction
Cloud Type	Observation	Name of Cloud Types
Present Weather	Observation	Clear, Rain, Partly Cloudy, etc.
7–12 Hour Forecast	Chart	Not applicable

QUESTIONS

1. How best might the cloud types be identified?
   (1) check the local newspaper
   (2) call the local weather station
   (3) compare with a chart of cloud types
   (4) guess
2. Set up a computer data base or spreadsheet that will allow club members to record weather information the same time each day.

(Continued)

3. The weather station thermometer measures temperature in degrees Fahrenheit. The Temperature Conversion Scale is used to convert Fahrenheit and Celsius temperatures. Explain briefly how to obtain the Celsius temperature if you know the Fahrenheit temperature.

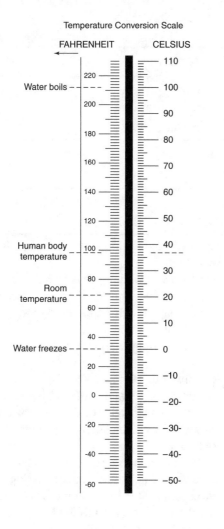

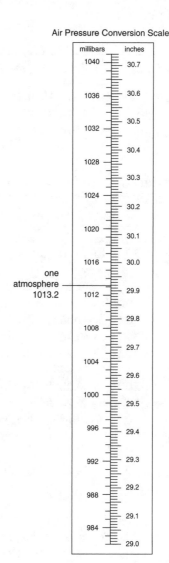

4. The weather station's recording barometer measures the air pressure in inches of mercury. The Air Pressure Conversion Scale can be used to convert inches of mercury to millibars. Determine the millibar reading for each of the inches of mercury readings:

(1) 29.92 inches of mercury
(2) 30.10 inches of mercury
(3) 29.75 inches of mercury

(Continued)

5. The weather predictions will be made using the Table: Weather Forecasting by Wind and Barometer. Locate the wind direction in the column on the left. Then find the best fit barometer reading in the center column. The weather forecast is found in the column on the right. Determine the forecast for each of the following conditions:

Winds	Barometer Reading	Tendency
1) Southwest	30.15	rising rapidly
2) South	29.75	falling rapidly
3) Southwest	29.95	rising slowly

Weather Forecasting by Wind and Barometer

Wind Direction	Barometer Reading	Forecast
SW to NW	30.10 to 30.20; steady	Fair, with slight temperature changes, for 1 to 2 days
SW to NW	30.10 to 30.20; rising rapidly	Fair, followed within 2 days by rain
SW to NW	30.20 and above; stationary	Continued fair, with no temperature change
SW to NW	30.20 and above; falling slowly	Slowly rising temperature and fair for 2 days
S to SE	30.10 to 30.20; falling slowly	Rain within 24 hours
S to SE	30.10 to 30.20; falling rapidly	Wind increasing, with rain within 12 to 24 hours
SE to NE	30.10 to 30.20; falling slowly	Rain in 12 to 18 hours
SE to NE	30.10 to 30.20; falling rapidly	Increasing wind; rain within 12 hours
E to NE	30.10 and above; falling slowly	In summer, with light winds, rain may not fall for several days. In winter, rain within 24 hours
E to NE	30.10 and above; falling rapidly	In summer, rain likely within 12 to 24 hours. In winter, rain or snow, with increasing winds, often starts when barometer begins to fall and the wind sets in from the NE.
SE to NE	30.00 or below; falling slowly	Rain will continue 1 to 2 days
SE to NE	30.00 or below; falling rapidly	Rain, with high wind, followed within 36 hours by clearing, in winter by colder temperatures
S to SW	30.00 or below; rising slowly	Clearing within a few hours; fair for several days
S to E	29.80 or below; falling rapidly	Severe storm imminent, followed within 24 hours by clearing, in winter by colder temperatures
E to N	29.80 or below; falling rapidly	Severe northeast gale and heavy precipitation; in winter, heavy snow followed by a cold wave
Going to W	29.80 or below; rising rapidly	Clearing and colder

Multiple Choice

18. During winter, air masses that form over northern Canada often affect the weather in New York State. Such an air mass would be

    (1) dry and warm   (3) moist and warm

    (2) dry and cool   (4) moist and cool

19. Air pressure within an air mass is usually

    (1) high in the center and low on the edges

    (2) high on the edges and low in the center

    (3) constant throughout

    (4) varying throughout

20. The major storm systems of our latitudes are

    (1) high-pressure systems   (3) tropical air masses

    (2) cold, dry air masses   (4) low-pressure systems

21. At which city in the diagram would the air pressure most likely be the greatest?

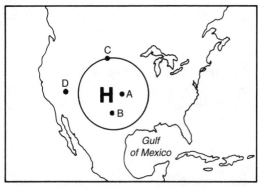

    (1) City *A*   (2) City *B*   (3) City *C*   (4) City *D*

22. Changes in air pressure indicate

    (1) the change of seasons

    (2) the passing of highs and lows

    (3) the climate is changing

    (4) the sun is setting

23. Margaret observed the weather from Friday night to Saturday night. First it was hot and humid, then there were thunderstorms, and finally the air became cooler and drier. These changes were probably due to the passing of a

    (1) warm front   (2) cold front   (3) hurricane   (4) wind belt

24. The diagram shows an air mass entering the United States from the northwest. This air mass formed over the North Pacific Ocean, so it would be

(1) moist and warm

(2) moist and cool

(3) dry and warm

(4) dry and cool

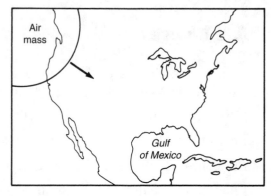

25. In the diagram, the air mass most likely to affect New York State the next day would be

(1) air mass *A*

(2) air mass *B*

(3) air mass *C*

(4) air mass *D*

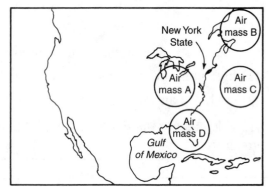

Questions 26 and 27 refer to the following diagram.

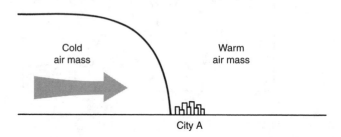

26. The line separating the cold air mass from the warm air mass represents a

(1) warm front   (3) line of latitude

(2) cold front   (4) high-pressure system

27. City *A* is most likely about to experience

(1) a light rain, followed by warmer temperatures

(2) heavy rains lasting several days

(3) no change in weather conditions

(4) brief downpours, followed by cooler temperatures

Thinking and Analyzing

Base your answers to questions 28 and 29 on the August weather map of New York State.

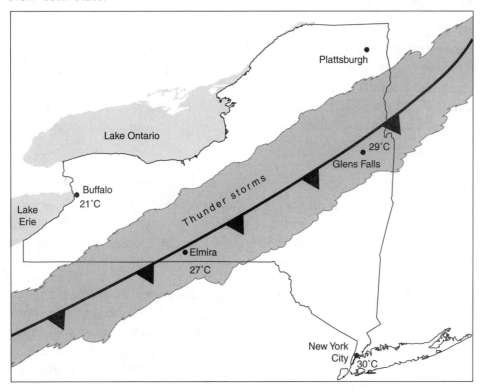

Present Weather Conditions

Buffalo—Cear skies and cool temperature

Glens Falls—windy and thunderstorms

Elmira—windy and thunderstorms

New York City—clear skies, hot and humid

28. Describe what the weather is most likely to be in Plattsburgh.

29. Predict the weather most likely to occur in New York City within the next 6 hours.

30. Describe the weather in a low-pressure system. Address the following weather conditions:

    a. vertical air motion

    b. horizontal air motion

    c. precipitation or lack of precipitation

    d. cloud cover or lack of cloud cover

31–32. On a weather map. Mark saw the word "High" over New York State.

31. What does "High" indicate on the map?

32. What type of weather would you expect to find in New York State?

Weather Hazards

Weather directly affects our lives. Lack of rain can ruin crops and force emergency measures such as water rationing. Too much rain can cause destructive floods. Storms produce severe weather conditions that can make travel dangerous, force schools and businesses to close, and cause deaths, injuries, and property damage. Given sufficient warning, humans can prepare for hazardous weather and minimize damage to life and property.

Storms

Storms are natural disturbances in the atmosphere that involve low air pressure, clouds, precipitation, and strong winds. The major types of storms are *thunderstorms, hurricanes, tornadoes,* and *winter storms*. Each type has unique characteristics and dangers.

1. ***Thunderstorms*** are brief (few hours), intense storms that affect a small area and are usually in the range of 10–20 km in size. They are produced when rapidly rising warm air causes cumulus clouds to build upward into a *thunderhead* (Figure 7-15). Thunderstorms are characterized by thunder and lightning, strong gusty winds, and sometimes hail.

   Lightning is a huge electrical discharge, like a giant spark. Lightning strikes are very dangerous and can be fatal. Large hailstones that sometimes fall during thunderstorms can also be dangerous. During a thunderstorm, you should stay indoors and especially avoid hilltops, open fields, beaches, and bodies of water.

Figure 7-15. *Development of a thunderhead.*

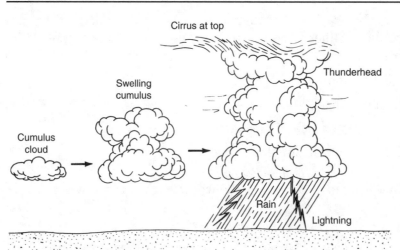

Figure 7-16. Hurricane Floyd striking the southeast coastline of the United States.

2. *Hurricanes* are huge, rotating storms that form over the ocean near the equator (Figure 7-16). Commonly, they are 700–800 km in diameter, and take about a day to pass. They produce very strong winds, heavy rains, and large, powerful waves. A calm region in the storm's center is called the *eye* of the hurricane.

   Hurricanes can cause severe flooding and damage from their high winds. People living along the coast and in flood-prone regions should leave their homes and move to higher ground when a hurricane strikes.

3. *Tornadoes* are violently whirling winds, sometimes visible as a funnel-shaped cloud (Figure 7-17). They are produced by particularly severe thunderstorms. Tornadoes usually appear suddenly, carve a narrow path of destruction, and disappear as suddenly as they

Figure 7-17. A tornado funnel extending from cloud to ground.

came. Spiraling high-speed winds and extremely low air pressure are the unique features of tornadoes.

A tornado can lift and toss large objects, including cars, into the air. It can destroy houses in its path in a matter of seconds. An underground cellar or basement is the safest place to be during a tornado.

4. **Winter storms** include blizzards and ice storms. *Blizzards* are fierce storms with strong winds, blowing snow, and very cold temperatures. *Ice storms* occur when falling rain freezes at Earth's surface, coating everything with ice. Under these conditions, you should remain indoors and not attempt to travel.

Process Skill

How Do Hurricanes Form and Travel in the North Atlantic Ocean?

North Atlantic hurricanes develop near the equator where the right weather conditions exist for their formation. Hurricanes need warm ocean water, warm, moist air, and low air pressure regions to develop. These conditions frequently occur in the Atlantic Ocean between 5° and 20° north of the equator from June to November.

A hurricane will maintain or grow in strength as long as these conditions are met. As the hurricane travels northward and/or over land some of these conditions are lost, and the storm begins to decay.

The map shows the path of Hurricane Floyd (1999). The storm originated about 14° N latitude in the Atlantic Ocean. As it was pushed westward by the trade winds it curved to the right and moved northward. At 30° N latitude it was affected by the westerlies, a wind belt that steers weather systems across the United States from west to east. At the middle latitudes the hurricane moved northeast and died.

Hurricane Floyd—September 7–17, 1999

DATE	LAT° N	LONG° W	WIND (mph)
09/07			25
09/08	17	52	45
09/09	18	57	60
09/10	21	60	75
09/11	23	64	95
09/12	23	68	110
09/13	24	74	135
09/14	27	77	120
09/15	31	79	100
09/16	39	75	55
09/17	43	70	50

(Continued)

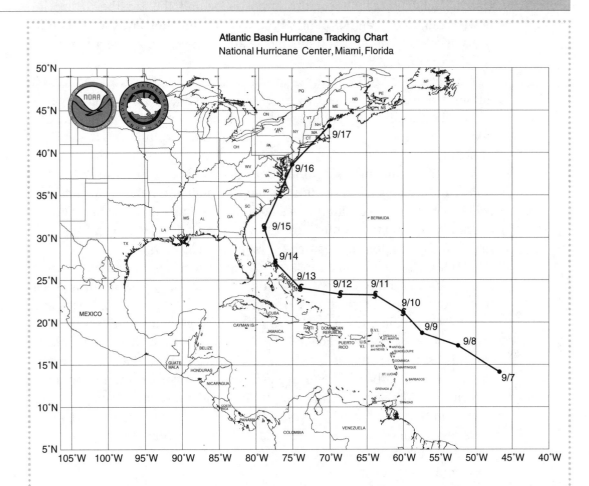

Atlantic Basin Hurricane Tracking Chart
National Hurricane Center, Miami, Florida

QUESTIONS

1. Hurricanes by definition must have sustained wind speeds of 74 miles per hour or greater. Tropical storms must have sustained wind speeds between 39 and 73 miles per hour. When the wind speed is less than 39 miles per hour, the storm is called a tropical depression.
   (a) How many days was Floyd a hurricane?
   (b) How many days was Floyd a tropical storm?
   (c) How many days was Floyd a tropical depression?
2. What was Floyd's latitude and longitude on September 7 (09/07)?
3. What is the most likely location of Floyd on September 18 (09/18)?
   (1) 39°N 74°W
   (2) 45°N 60°W
   (3) 43°N 75°W
   (4) 40°N 70°W
4. Between September 15 and 17 Hurricane Floyd decreased in wind speed considerably. Why did this occur?
5. On what day was the hurricane warning for Miami discontinued?

Pollution

Human activities can affect the atmosphere and the weather. Factories, power plants, cars, and airplanes produce harmful substances called **pollutants**. The build-up of pollutants in the atmosphere can cause a number of weather problems.

Smog is a haze formed by the reaction of sunlight with chemicals in automobile exhaust and factory smoke. Smog tends to hang over large cities, giving the air a hazy, dirty look. Inhaling smog is very dangerous for people with breathing problems like asthma, and harmful even to the lungs of healthy people.

Chemicals in smoke from factories and vehicles can also increase the *acidity* of the moisture in clouds. When this moisture falls to Earth as **acid rain**, it can harm lakes and forests, and the creatures that live in them. Because air pollutants are often carried along by the prevailing winds, acid rain may fall far from the source of pollution (Figure 7-18).

Many industries and most forms of transportation produce *carbon dioxide*. In the atmosphere, carbon dioxide acts like the glass of a greenhouse, trapping heat close to Earth instead of letting it radiate back into space. This is called the **greenhouse effect**. The greenhouse effect plays an important natural role in keeping Earth warm. Without it Earth would be too cold to support life as we know it. However, many scientists fear that the buildup of carbon dioxide in the atmosphere caused by human activities may lead to global warming, a rise in worldwide average temperatures. This condition, often called a *runaway greenhouse effect*, could have disastrous results, making climates hotter and drier, and interfering with agriculture. Polar ice caps could melt, raising the sea level and flooding coastal areas and many major cities.

Certain natural events also release pollutants. Forest fires and volcanic eruptions give off huge quantities of dust and ash particles that collect high in the atmosphere, blocking sunlight and causing cooler tem-

Figure 7-18. Acid rain may affect areas far downwind from the source of pollution.

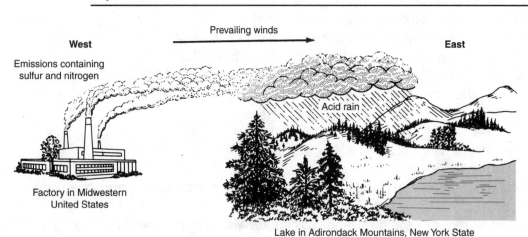

Prevailing winds

West

Emissions containing sulfur and nitrogen

East

Acid rain

Factory in Midwestern United States

Lake in Adirondack Mountains, New York State

peratures on Earth. Plants release irritating pollen into the air, causing health problems for people with hay fever and asthma. There is little that we can do to control such natural pollutants. However, we can, and must, control our own activities that cause pollution if we are to avoid harming our planet

Process Skill

Interpreting Data in a Table

Storms form when the temperature, air pressure, and moisture conditions necessary for their development exist. At certain times of the year, weather conditions are more likely to produce one of the three types of storms—thunderstorms, tornadoes, and hurricanes—than at other times. Therefore, each storm type should have a "storm season."

You can determine if there are "storm seasons" by keeping track of the storms of each type that strike during the year, and recording which months they occur in, as shown in the table below. With this information, you can see if each kind of storm takes place mostly within a definite time of year—a "storm season."

Number of Storms Per Month

Month	Thunderstorm	Tornado	Hurricane	Month	Thunderstorm	Tornado	Hurricane
January	0	1	0	July	18	2	1
February	1	0	0	August	16	1	5
March	3	2	0	September	10	1	8
April	6	4	0	October	6	0	4
May	9	5	0	November	2	1	1
June	15	4	0	December	0	1	0

According to the table, most thunderstorms occur in the months of June, July, and August, so there does appear to be a thunderstorm season. Is there evidence in the table of a season for tornadoes? How about for hurricanes? Use the table to answer the following questions.

QUESTIONS

1. During which three months do most tornadoes occur?
   (1) August, September, and October   (3) June, July, and August
   (2) April, May, and June   (4) May, June, and July
2. Which storm type has the most clearly defined "storm season"?
   (1) thunderstorm   (2) tornado   (3) hurricane
   (4) all three types have equally well-defined seasons

Multiple Choice

33. The main hazard of a thunderstorm is

(1) thunder

(3) funnel-shaped winds

(2) heavy rains

(4) lightning

34. The violent windstorm visible as a funnel-shaped cloud in the illustration is called a

(1) hurricane

(3) tornado

(2) thunderstorm

(4) blizzard

35. A hurricane is approaching the east coast of Florida. What dangers should the people there take precautions against?

(1) cold temperatures, blowing snow, and poor visibility

(2) funnel-shaped winds that can lift large objects

(3) lightning and hailstones

(4) flooding, large waves, and strong winds

36. Substances released into the atmosphere by the activities of humans

(1) always have a positive effect on the weather

(2) can have a harmful effect on the weather

(3) have no effect on weather

(4) may cause global cooling

37. John visited the city on a warm summer day. He noticed that the air was hazy, and that his eyes and throat burned. This was probably caused by

(1) an approaching storm

(3) low clouds

(2) smog

(4) global warming

38. When scientists speak of the "runaway greenhouse effect," they are referring to

(1) the fact that vegetables grown in a greenhouse do not taste as good as vegetables grown outdoors

(2) the use of green paint on houses to camouflage them

(3) the idea that pollution caused by human activities may lead to global warming

(4) the fact that ash from volcanic eruptions can cause cooler temperatures

39. The map shows a major industrial city and three lakes in the central U.S. Which lake is most likely to be affected by acid rain caused by pollution from the city?

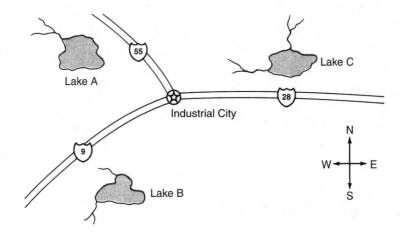

(1) lake *A*

(2) lake *B*

(3) lake *C*

(4) all three lakes will be affected equally

40. The table below lists sources of air pollutants and the percentage contributed by each source.

Sources and Percentages of Air Pollutants

Pollutant Source	Percentage of Total Pollutants
Transportation	42%
Fuel	21%
Solid waste disposal	5%
Forest fires	8%
Miscellaneous	10%
Industrial processes	14%

Which pie graph correctly represents the data in the table?

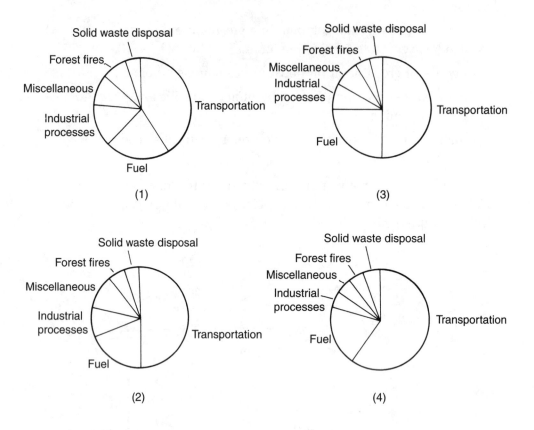

(1)

(2)

(3)

(4)

Thinking and Analyzing

The following diagram and graph show the cross section of a typical hurricane and the change in wind speed within the hurricane. Carefully review the diagram and compare it to the graph, and answer questions 41 to 44.

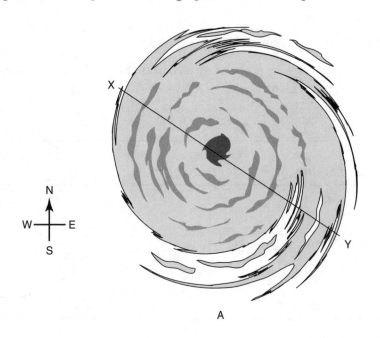

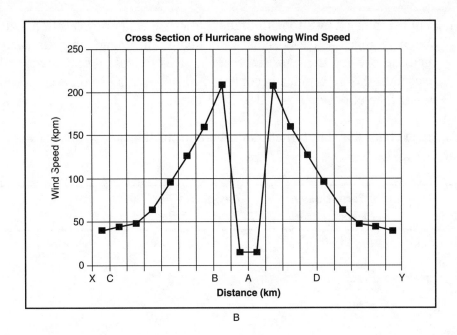

B

41. If this hurricane is 720 km across what is the distance between each of the marks on the distance axis line in Figure B?

42. What point on the distance axis line represents the center of the hurricane?

43. The eye of a hurricane is an area of generally calm winds and clear skies. What is the diameter of the eye of this hurricane?

44. Describe the relationship of the winds to the structure of a hurricane.

Questions 45 and 46 are based on the diagram, which shows the three stages in the development of a thunderstorm.

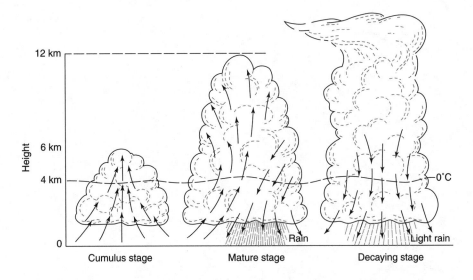

45. Above what elevation does the cloud contain ice crystals rather than water droplets?

46. Hail is produced by strong updrafts that cause water droplets to freeze. When falling, the ice crystals accumulate water, and updrafts carry them up causing them to freeze again. This process repeats over and over until the ice crystals become too heavy and fall to the ground. In what stage of development does hail form?

Using the chart showing tornado information for five states, answer questions 47 to 49.

State	# Tornadoes in 25-Year Period	# Tornadoes per 10,000 sq. miles
New York	93	0.7
Oklahoma	1401	7.7
Texas	3035	4.4
Florida	1006	6.6
California	84	0.2

47. In which state is a house least likely to be hit by a tornado?

48. What might explain the fact that Texas has many more tornadoes than Oklahoma, but a house is in more jeopardy to be hit by a tornado in Oklahoma?

49. About how many tornadoes per year occur in New York State?

Chapter 8
Astronomy

Points to Remember

▶ Rotation is the spinning of Earth on its axis, which causes day and night. It takes Earth 24 hours (one day) to complete one rotation.

▶ Revolution is Earth's motion around the sun; it takes Earth 365.25 days (one year) to complete one revolution around the sun.

▶ Earth's axis is tilted 23.5° to the plane of its orbit. The northern end of Earth's axis (the North Pole) points toward the North Star. The seasons are caused by the tilt of Earth's axis as it revolves around the sun. This tilt causes the sun's vertical rays to strike Earth farthest north on June 21 and farthest south on December 21.

▶ Our solar system consists of the sun, the nine planets, satellites, asteroids, meteoroids, and comets. The sun is at the center; its gravity keeps all other members of the solar system in orbits around the sun.

▶ The universe contains trillions of stars most clustered together in galaxies. Yet most of the universe is considered to be empty space. Our sun is an average star, one of 800 billion stars clustered in a spiral galaxy called the Milky Way.

Earth Motions, Time, and Seasons

Earth's Rotation

People once believed that Earth stood still while the sun, moon, stars, and planets revolved around Earth each day. This seems reasonable, since we do not feel Earth moving, and the sun does appear to move across the sky during the day while the moon and stars appear to move at night. However, today we know that these apparent motions of the sun, moon, and stars are actually caused by Earth's rotation.

Earth spins like a top. This spinning motion is called **rotation**. Earth rotates from west to east, or, put another way, if we could look down at the North Pole from space, we would see Earth spinning in a counterclockwise direction.

Extending through Earth between the North and South Poles is an imaginary rod, or **axis of rotation**, on which Earth spins (see Figure 8-1). A basketball spinning on a fingertip gives a good idea of how Earth spins on its axis. The line from the fingertip through the basketball to the top of the ball is the axis of rotation.

Earth's rotation produces several effects:

1. Earth's rotation causes the daily change from day to night. At any given time, half of Earth is in daylight, facing the sun, while half is in darkness, facing away from the sun. This is shown in Figure 8-2. Every day, all places on Earth, except the areas near the poles, experience this change from daylight to darkness. (Areas within the Arctic and Antarctic circles experience several weeks of continuous daylight or darkness at certain times of the year.)

Figure 8-1. *Rotation of earth: (a) Earth rotates from west to east around its axis. (b) Viewed from above the North Pole (NP), Earth rotates in a counterclockwise direction.*

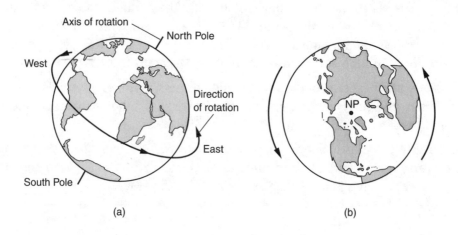

(a) (b)

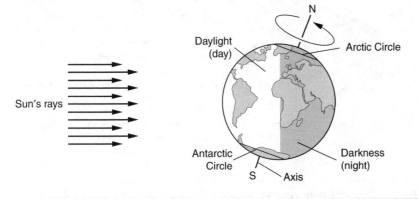

Figure 8-2. *Earth's rotation cause the change from day to night.*

2. The speed of Earth's rotation causes the length of one day to be about 24 hours. This is the amount of time Earth takes to rotate once on its axis.
3. The apparent daily motions of the sun, moon, planets, and stars across the sky are also caused by Earth's rotation. These objects appear to rise in the eastern sky and set in the western sky because Earth rotates from west to east.

Daily Time

The time of day at any location on Earth is based on Earth's rotation and the position of the sun in the sky. Earth completes one rotation each day. That is, every location on Earth rotates 360° around Earth's axis each day. There are 24 hours in a day. If we divide 360° by 24 hours, we find that Earth rotates 15° per hour. Table 8-1 and Figure 8-3 on page 170 show how the time of day and the location of the sun in the sky are related to Earth's rotation.

Earth's *longitude*, imaginary lines that run north and south, is measured in degrees east, and west, and is based on the rotation of Earth. Every 15°-change in longitude represents a difference of one hour of Earth time.

Table 8-1. *Rotation Positions on Earth*

Line	Rotation Angle	Time	Location of Sun
A-NP	Start	12:00 noon	High in the sky
B-NP	45°	3:00 P.M.	In the western sky
C-NP	90°	6:00 P.M.	Near western horizon
D-NP	180°	12:00 midnight	Opposite side of Earth
E-NP	270°	6:00 A.M.	Near eastern horizon

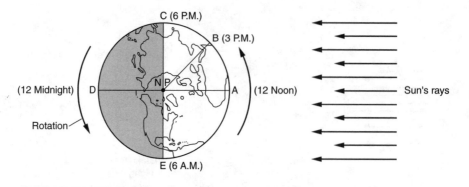

Figure 8-3. *The time of day and the sun's location in the sky are related to Earth's rotation. (That is, it is 6 A.M. everywhere on line NP-E.)*

Earth's Revolution

Earth moves around the sun in a motion called **revolution**. The path Earth travels around the sun is called an **orbit**. Earth's orbit is not perfectly circular, it is actually slightly oval in shape (Figure 8-4).

Earth's revolution has two major effects. First, the time Earth takes to revolve once around the sun defines the length of a year. During that time, Earth rotates on its axis $365\frac{1}{4}$ times, so there are $365\frac{1}{4}$ days in a year. For convenience, the calendar year is 365 days long, and an extra day is added every fourth year (called a leap year) to make up for each leftover $\frac{1}{4}$ day.

Second, Earth's revolution around the sun, combined with the tilt of Earth's axis, causes the changing seasons on the planet. Earth's axis of rotation is not perpendicular (at a right angle) to the plane of its orbit; rather, it is tilted 23.5° (see Figure 8-5).

No matter where Earth is in its orbit, its axis is always tilted in the same direction in space, pointing toward the North Star. While all the other stars seem to move across the night sky, the North Star remains motionless because Earth's axis points toward it.

Figure 8-4. *Earth's orbit around the sun is oval, not a perfect circle.*

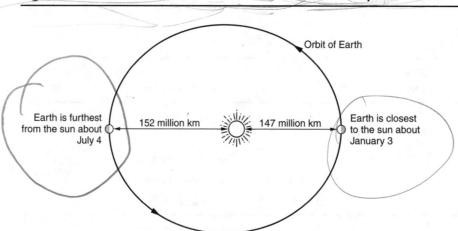

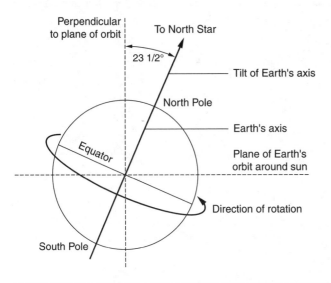

Figure 8-5. *Earth's axis of rotation is tilted 23½°.*

Yearly Seasonal Time

The length of a year is based on Earth's revolution. Earth revolves completely around the sun once each year. That is, Earth revolves 360° during a 12-month period of time. If we divide 360° by 12 months, we find that Earth revolves about 30° per month. Table 8-2 and Figure 8-6 on page 172 show how the date and the season on Earth are related to Earth's revolution around the sun.

Seasonal Changes

Earth's orbit takes it closest to the sun in early January and farthest from the sun in early July (refer to Figure 8-4). This means it is not Earth's changing distance from the sun that causes the changing seasons. The cause is the tilt of Earth's axis as the planet revolves around the sun. Because the axis always points in the same direction while Earth orbits the sun, the Northern Hemisphere is tilted toward the sun for half the year and away

Table 8-2. *Revolution Positions of Earth*

Point	Revolution Angle	Date	Seasonal Information
A	Start	December 21	Winter begins
B	30°	January 21	Winter
C	90°	March 21	Spring begins
D	180°	June 21	Summer begins
E	270°	September 23	Autumn begins

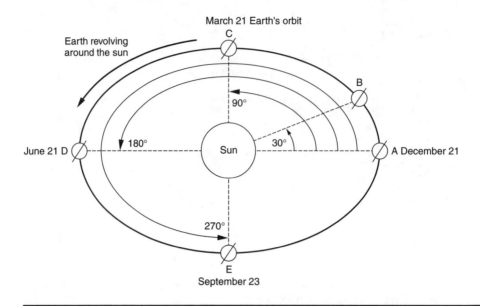

Figure 8-6. *Earth revolves 360° around the sun in a 12-month period.*

from the sun the other half of the year (see Figure 8-7). This causes changes in the number of hours of daylight each day and in the angle at which the sun's rays strike Earth.

The sun's vertical rays strike Earth when the sun is directly overhead. The vertical rays are the strongest and hottest rays, and are therefore significant in causing the seasons on Earth.

On June 21, the Northern Hemisphere is tilted toward the sun, and the sun's vertical rays reach 23.5° north of the equator, the Tropic of Cancer (see

Figure 8-7. *The seasons are caused by the tilt of Earth's axis and Earth's revolution around the sun.*

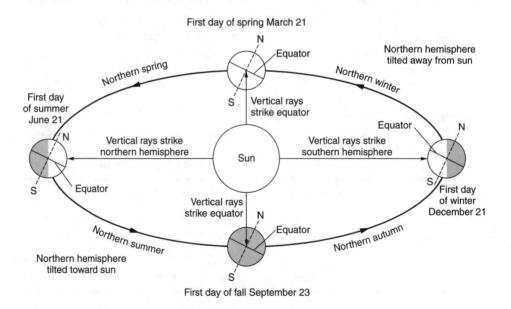

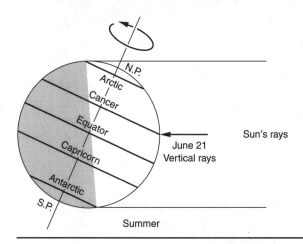

Figure 8-8. *On June 21, the sun's vertical rays strike the Tropic of Cancer.*

Figure 8-8). The Northern Hemisphere is heated more effectively and begins summer. On the first day of summer in the Northern Hemisphere, the sun follows its highest path across the sky (see Figure 8-9). The area within the Arctic Circle has 24 hours of daylight, and as you travel south the number of daylight hours decreases until you reach the Antarctic Circle. Everywhere within the Antarctic Circle has 24 hours of darkness (night).

On December 21 the Northern Hemisphere is tilted away from the sun, and the sun's vertical rays reach 23.5° south of the equator, the Tropic of Capricorn (see Figure 8-10 on page 174). It is the first day of winter in the Northern Hemisphere, the sun follows its lowest path across the sky (see again Figure 8-9). The area within the Arctic Circle has 24 hours of darkness, and as you travel south the number of daylight hours increases until you reach the Antarctic Circle. Everywhere within the Antarctic Circle has 24 hours of daylight.

Figure 8-9. *The sun's apparent path across the sky changes with the seasons.*

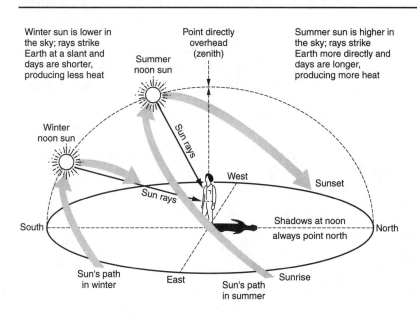

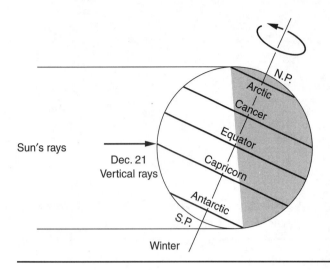

Figure 8-10. On December 21, the sun's vertical rays strike the Tropic of Capricorn.

In the Southern Hemisphere, the situation is reversed. During summer in the Northern Hemisphere, the Southern Hemisphere experiences winter. During winter in the Northern Hemisphere, the Southern Hemisphere has its summer.

On March 21 and September 23 neither hemisphere is really tilted toward the sun. The sun's vertical rays strike the equator, and all areas of Earth have equal periods of daylight and darkness (see Figure 8-11). Table 8-3 summarizes information about seasonal dates in the Northern Hemisphere.

During any season, the sun is highest in the sky each day at noon. In the continental United States, the noon sun approaches but never reaches the point directly overhead, it is always in the southern half of the sky (see again Figure 8-9). As a result, shadows at noon always point to the north.

Proof of Earth's Revolution

Proof that Earth revolves around the sun comes from observations of stars. Stars in the night sky form patterns that have reminded people of animals

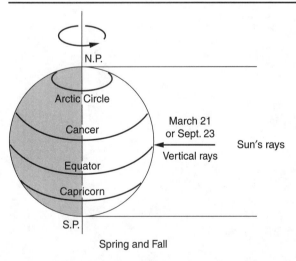

Figure 8-11. On March 24 and September 23, the sun's vertical rays strike the equator.

Table 8-3. *Seasonal Information for the Northern Hemisphere*

Date	Season	Sun's Path Across the Sky	Location of Vertical Rays	Length of Daylight and Darkness per Day
June 20 or 21	First day of summer	Sun follows highest path across sky	Tropic of Cancer	Longest period of daylight; shortest period of darkness
September 22 or 23	First day of autumn	Sun follows path midway between summer and winter extremes	Equator	Daylight and darkness of equal length (12 hours each)
December 21 or 22	First day of winter	Sun follows lowest path across sky	Tropic of Capricorn	Shortest period of daylight; longest period of darkness
March 20 or 21	First day of spring	Sun follows path midway between summer and winter extremes	Equator	Daylight and darkness of equal length (12 hours each)

or characters in ancient myths. These patterns are called **constellations**. Two easily recognized constellations are Ursa Major (the Great Bear, which contains the Big Dipper) and Orion, the Hunter.

During the course of the year, different constellations become visible at night. This suggests that Earth's night side faces different directions in space (Figure 8-12), so Earth's position in relation to the sun must be changing. In other words, Earth must be moving around the sun.

Figure 8-12. *As Earth orbits the sun, different constellations become visible at night.*

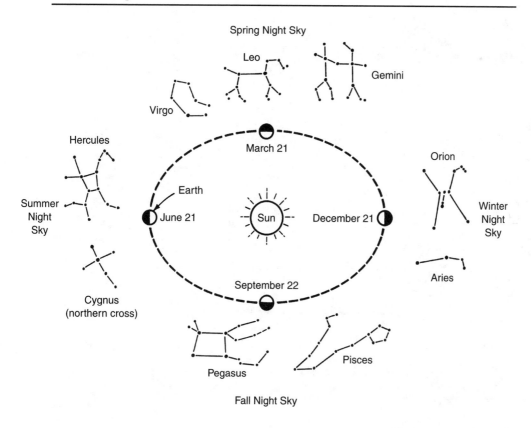

Designing a Measurement Procedure

As you have learned, in the United States the noon sun is not directly overhead, but is always in the southern sky. However, the noon sun is much higher overhead in summer than in winter. In fact, the sun's position at noon changes a little each day, moving higher in the sky from December 21 to June 21, and lower in the sky from June 21 to December 21. As the height of the sun changes, so does the angle of its rays striking Earth. At any location on Earth, the angle at which the sun's rays strike Earth's surface at noon changes from day to day.

You can determine the changing angle of the sun's rays by measuring the shadow of a vertical pole. Because the noon sun is always in the southern sky, the shadow of the pole is always cast to the north. The length of the shadow each day indicates the changing position of the noon sun. As the noon sun moves higher in the sky, the angle of its rays becomes larger and the length of the shadow gets shorter. As the sun moves lower, the angle of its rays becomes smaller and the shadow gets longer. This is shown in Diagram 1.

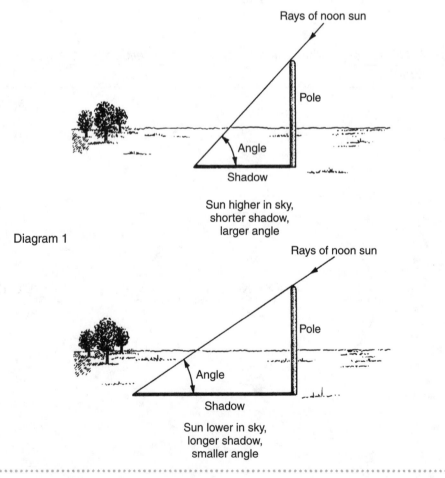

Diagram 1

(Continued)

Kent wanted to measure the changing angle of the noon sun's rays near Syracuse, New York. He set up an experiment using a vertical pole and a tape measure that ran north along the ground from the base of the pole, as shown in Diagram 2. For eight weeks, Kent measured the length of the pole's shadow every Friday at 12:00 noon. Study the diagrams and then answer the following questions.

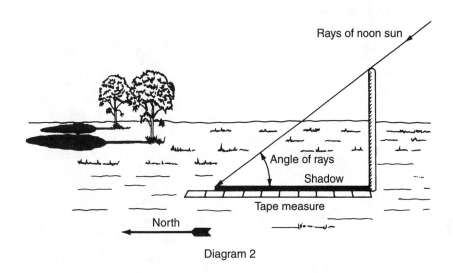

Diagram 2

QUESTIONS

1. If the length of the shadow increased each week during the experiment, this would indicate that the noon sun was
 (1) getting higher in the sky, producing a larger angle
 (2) getting higher in the sky, producing a smaller angle
 (3) getting lower in the sky, producing a larger angle
 (4) getting lower in the sky, producing a smaller angle

2. If Kent had started the experiment on December 21 and obtained a shadow length of 51 centimeters, describe what the length of the shadow would be on January 15. Explain why this would occur.

3. If Kent had started the experiment on May 21 and continued his experiment until July 21, describe how the shadow would change, and explain why.

Review Questions

Multiple Choice

Questions 1 and 2 refer to the map below, which shows New York State with about half the state in daylight and half the state in darkness (night). As Earth rotates from west to east, the areas of daylight and night in the state will change.

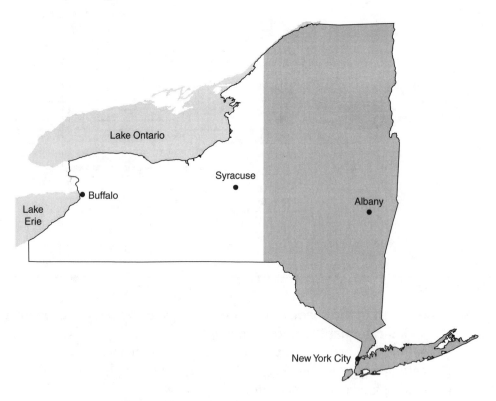

1. The time of day in Syracuse is most nearly

    (1) noon   (3) midnight

    (2) dusk (sunset)   (4) dawn (sunrise)

2. As Earth rotates from west to east over the next few hours, New York State will

    (1) move completely into night

    (2) move completely into daylight

    (3) remain unchanged in terms of its areas of daylight or darkness

    (4) change from being completely night to completely daylight

3. The sun rises in the east and sets in the west because Earth rotates from

    (1) west to east   (3) north to south

    (2) east to west   (4) north to west

4. The daily change from daylight to darkness is caused by the

(1) Earth's revolution around the sun

(2) tilt of Earth's axis

(3) Earth's rotation on its axis

(4) sun's light going out at night

5. Earth rotates once on its axis in one complete

(1) year   (3) month

(2) week   (4) day

As Earth revolves around the sun each year, the sun's vertical rays move from 23.5° south of the equator to 23.5° north of the equator and back again. This is due to the combined effect of Earth's revolution around the sun and the tilt of Earth's axis. Questions 6 to 7 refer to the diagram below.

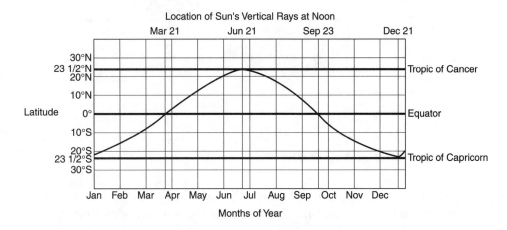

6. New York extends between 41° and 45° north latitude. The sun's vertical rays strike New York

(1) between March 21 and June 21

(2) between June 21 and September 23

(3) June 21

(4) never

7. The sun's vertical rays strike the equator on

(1) June 21   (3) March 21 and September 23

(2) December 21 and June 21   (4) December 21

8. During the summer months (June 21 to September 23), the location on Earth that receives the sun's vertical rays is

(1) farthest south

(2) changing from north to south

(3) changing from south to north

(4) not changing

9. Earth makes one complete revolution around the sun in one

    (1) year   (3) month

    (2) week   (4) day

10. Tracy read in the newspaper that there would be exactly 12 hours of daylight and 12 hours of darkness the next day. That next day's date most likely is

    (1) October 21   (3) December 21

    (2) June 21   (4) March 21

11. At noon one day, Bob noticed the shadow of a tree in Binghamton, New York. What direction is Bob facing?

    (1) north

    (2) east

    (3) south

    (4) west

Thinking and Analyzing

12. Jason saw the moon over a tree at 9:00 P.M. (position A in the diagram). An hour later, the moon had moved to position B. What caused this apparent change in position, and where will the moon be in another hour?

Questions 13 and 14 refer to the following diagram. Earth revolves around the sun in a nearly circular orbit. In one year, Earth makes one complete revolution around the sun. Therefore, during its $365\frac{1}{4}$ day trip around the sun, it completes a 360° orbit. That means that Earth revolves at a rate of about 1° per day.

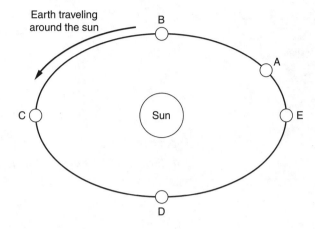

13. During one month (30 days), about how far will Earth revolve around the sun?

14. On January 1, Earth is located at position E in the diagram. At what point in its orbit would Earth be three months later, and nine months later?

15. Earth is closer to the sun on December 21 than on June 21. What conclusion about seasons can be made from this information?

The Solar System

The *solar system* consists of our sun and all the objects that revolve around it. The major members of the solar system are the sun and the nine *planets* (see Figure 8-13). A number of other objects also come under the influence of the sun's gravity, and so belong to the solar system. These include satellites, or moons (objects that revolve around planets), asteroids, comets, and

Figure 8-13. *The solar system.*

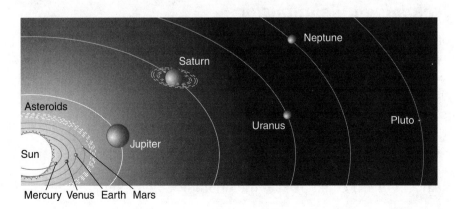

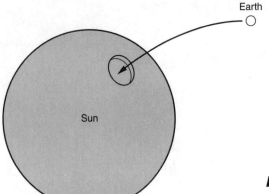

Earth

Sun

Figure 8-14. *If the sun were hollow, more than a million Earths could fit in it.*

meteoroids. Most objects in the solar system have regular and predictable motion.

The Sun

The sun is a hot, bright ball of gases. Nuclear reactions in the sun's interior release enormous amounts of energy, mostly as light and heat. The sun is by far the largest object in the solar system. It is many times larger than Earth. If the sun were hollow, about one million Earths could fit inside. (See Figure 8-14.)

The sun is a *star*, like the stars we see at night. Although the sun is an average size star, it seems much larger than other stars because it is much closer to Earth. Light from the sun takes 8 minutes and 20 seconds to reach Earth. In contrast, light from the nearest star (other than the sun) takes more than four years to reach Earth! The sun is Earth's main source of energy, providing the light and heat necessary for the existence of life.

 *Process Skill*

Explaining a Relationship

In 1610, Galileo discovered moons, or satellites, revolving around other planets in the solar system. Using a telescope to observe the night sky, he saw four moons revolving around Jupiter. Other discoveries made since then include three previously unknown planets and many satellites.

Inquisitive astronomers discovered these three planets and some of the satellites. Johannes Kepler formulated laws governing the motion of objects in space. When objects were observed to move differently than predicted by these laws, astronomers investigated the cause. They found that the gravitational force exerted by some as yet unknown object could explain the unpredictable motions. Using

(Continued)

mathematical calculations, the approximate location and size of the object was determined. Searching for these objects led to the discovery of the three outermost planets and some of the satellites in the solar system.

The use of telescopes and space probes also aided in the discovery of distant satellites. The telescope was first used for astronomy in 1609, when Galileo made some initial discoveries with it. Today, after many technological improvements, different types of telescopes continue to look farther into space. Over the last 30 years, space probes sent to the outer reaches of the solar system have helped scientists discover many additional satellites.

The table below lists the known planets and the number of known satellites at different dates. Use the information presented above and in the table to answer the following questions.

Number of Known Satellites

Planet	1609	1892	1962	1982	2000
Mercury	0	0	0	0	0
Venus	0	0	0	0	0
Earth	1	1	1	1	1
Mars	0	2	2	2	2
Jupiter	3	4	12	16	17
Saturn	0	8	9	16	18
Uranus	ND*	4	5	5	20
Neptune	ND*	1	2	2	8
Pluto	ND*	ND*	0	1	1

*ND: Planet not discovered

QUESTIONS

1. The discoveries of additional planets and satellites over the last 300 years were made possible by
   (1) inquisitive scientists and improved technology
   (2) Kepler's formulation of the laws of planetary motion
   (3) the use of mathematical calculations
   (4) all of the above
2. Why has the number of known satellites for the inner planets (Mercury, Venus, Earth, and Mars) remained relatively constant, while the number of known satellites for the outer planets has increased over time?
3. Astronomers will most likely continue to find additional satellites around the outer planets because
   (1) technology keeps improving
   (2) mathematical skills are improving
   (3) their predictions are improving
   (4) the laws of motion are changing

The Planets

There are nine planets that revolve around the sun. In the night sky, the planets look much like stars. However, as days and weeks go by, planets change position against the background of motionless stars. Also, they tend not to "twinkle" the way stars do. Unlike stars, planets do not give off their own light. The planets are visible because they reflect the light of the sun.

Each planet has its own special characteristics. Mercury is the closest planet to the sun. Venus is the hottest planet and the brightest as seen from Earth. Mars is often called the Red Planet, because it reflects reddish light. Jupiter, the largest planet, is characterized by a huge red spot. Rings made of rock and ice particles encircle Saturn. Uranus' axis is tilted at such a great angle that it almost points to the sun. This extreme tilt makes the planet appear to roll along its orbit. Neptune is surrounded by bright blue clouds of frozen methane. Pluto is the smallest planet and usually the farthest from the sun, although part of Pluto's orbit crosses inside Neptune's orbit.

Earth, the third planet from the sun, is unique in that it appears to be the only planet in the solar system that supports life. Much of what we know about the other planets has been discovered with the help of space probes. Table 8-4 lists the planets and some important facts about them.

Table 8-4. Planetary Data

Planet (in order from Sun)	Distance from Sun (in Earth-Sun Units*)	Time to Revolve Once Around Sun	Time to Rotate Once on Axis	Number of Satellites (Moons)
Inner Planets				
Mercury	0.4	88 days	59 days	0
Venus	0.7	225 days	243 days	0
Earth	1.0	365.25 days	24 hours	1
Mars	1.5	1.88 years	24.6 hours	2
Asteroid belt				
Outer Planets				
Jupiter	5.2	11.86 years	9.9 hours	17**
Saturn	9.5	29.63 years	10.6 hours	18**
Uranus	19.2	83.97 years	17 hours	20**
Neptune	30.1	165 years	16 hours	8**
Pluto	39.5	248 years	6.4 days	1

*An Earth-sun unit is the average distance from Earth to the sun (149,600,000 kilometers).
**Number of known moons; the actual number may be higher.

Determining a Quantitative Relationship

For a planet to remain in orbit around the sun, the gravitational pull of the sun on the planet must be in balance with the planet's speed and distance from the sun. Without this balancing, the planet would not stay in orbit.

Since the sun's gravity becomes stronger the closer an object gets to it, planets near the sun have to move faster than planets farther away to avoid being pulled into the sun. This suggests that there is a relationship between a planet's orbital speed and its distance from the sun.

Planetary Data Table

Planet	Average Orbital Speed (kilometers per second)	Average Distance from the Sun (kilometers)
Mercury	47.60	57,900,000
Venus	34.82	108,200,000
Earth	29.62	149,600,000
Mars	23.98	227,900,000
Jupiter	12.99	778,000,000
Saturn	9.58	1,427,000,000
Uranus	6.77	2,871,000,000
Neptune	5.41	4,497,000,000
Pluto	4.72	5,913,000,000

The table reveals that as a planet's distance from the sun increases, its orbital speed decreases. This is called an inverse relationship, because as one quantity increases (distance from the sun), the other quantity decreases (orbital speed). Use the information presented above and in the data table to answer the following questions.

QUESTIONS

1. If Earth moved closer to the sun, to remain in orbit its orbital speed would have to _____
2. If a planet farther from the sun than Pluto were discovered, its orbital speed would probably be
   (1) faster than Pluto's
   (2) slower than Pluto's
   (3) the same as Pluto's
   (4) impossible to determine
3. If the orbital speed of a newly discovered asteroid were found to be 18.61 km/s, between what two planets would the asteroid's orbit be located?

The Moon

The moon is a ball of rock that revolves around Earth as Earth revolves around the sun. It is Earth's only natural satellite and our nearest neighbor in space. There is no water or air on the moon, so it cannot support life. Because the moon is much smaller than Earth, the moon's gravity is too weak to hold moisture or an atmosphere at its surface.

The moon has a variety of surface features. The dark areas of the moon are low, flat plains. The light areas are mountainous highlands. As shown in Figure 8-15, the moon's surface is pockmarked by numerous craters, which are circular pits ringed by walls, or rims. Rock fragments that struck the moon's surface formed most of the craters.

Motions and Phases of the Moon

The moon takes $29\frac{1}{2}$ days, about one month, to revolve around Earth from one full moon to the next (see Figure 8-16). As the moon orbits Earth, it also rotates on its axis. Because the moon completes one rotation in the same amount of time it takes to orbit Earth once, the same side of the moon always

Figure 8-15. *In this photograph of the moon's surface, darker areas are maria, and brighter areas are highlands.*

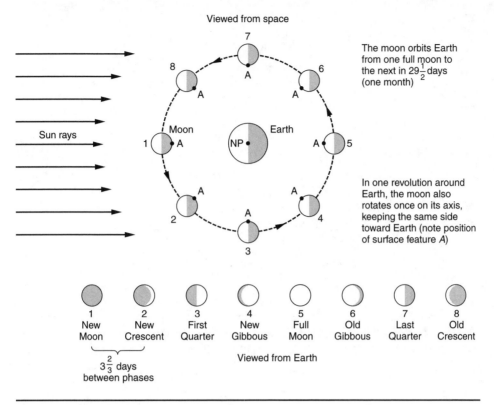

Figure 8-16. *Phases of the moon: The moon takes 29½ days to complete the cycle of its phases.*

faces Earth. Not until 1966, when a spacecraft circled the moon and took photographs, did we learn what the moon's far side looks like.

The moon does not give off its own light, but reflects light from the sun. The sun lights half of the moon's surface at all times, just as it always lights half of Earth. As the moon revolves around Earth, we see varying amounts of its lighted side, so the shape of the moon appears to change. These apparent changes in shape are called the **phases of the moon**.

At the phase called **new moon**, the moon is between Earth and the sun. The side facing Earth is dark, so the moon cannot be seen. At **full moon** phase, the moon is on the opposite side of Earth from the sun, and the side facing Earth is completely illuminated. Figure 8-16 also shows the major phases and their relationship to the moon's revolution around Earth. The moon takes about $3\frac{2}{3}$ days to move between each phase.

Eclipses

Events called *eclipses* sometimes occur when the sun, Earth, and moon are lined up. A *lunar eclipse* takes place when the moon passes through Earth's shadow. This can happen only when the moon is

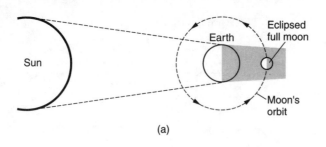

(a)

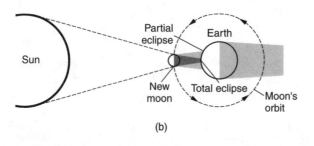

(b)

Figure 8-17. *Eclipses: (a) a lunar eclipse occurs when the moon passes into Earth's shadow; (b) a solar eclipse occurs when the moon casts its shadow is on Earth.*

full, as shown in Figure 8-17a. A lunar eclipse is visible over a large area on Earth and takes a few hours from start to finish.

A *solar eclipse* takes place when the moon casts its shadow on Earth, which can occur only during new moon phase, as shown in Figure 8-17b. A solar eclipse is visible over a small area on Earth. A *total eclipse*, in which the sun is completely blocked by the moon, lasts only a few minutes.

The Moon and Earth's Tides

The **tides** are the regular rise and fall in the level of ocean waters that take place twice each day. These changes in sea level are caused by the gravitational pull of the moon and, to a lesser extent, the sun. The moon is much closer to Earth than the sun is, so the moon's gravity affects Earth's tides more strongly than does the sun's gravity.

The pull of the moon's gravity draws the ocean waters into two large bulges, one on the side of Earth facing the moon and one on the opposite side of Earth. High tides occur at each of these two positions (Figure 8-18). Halfway between the tidal bulges, and one-quarter of the way around Earth, the ocean level falls, producing low tides. As Earth rotates, locations on Earth experience changing tides. The gradual change from high tide to low tide takes about $6\frac{1}{4}$ hours, and another $6\frac{1}{4}$ hours for the water to rise to high tide again.

When the sun, Earth, and moon are lined up, the gravity of the sun and moon combine to produce especially high and low tides. When they are not lined up, the tides are less extreme.

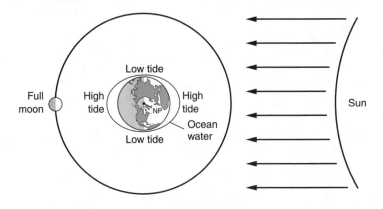

Figure 8-18. *High tides occur at positions on Earth facing directly toward or away from the moon. Low tides occur halfway between high tides.*

Asteroids, Comets, and Meteoroids

Asteroids are rock fragments of various shapes and sizes that revolve around the sun. Many are found in a region called the asteroid belt, between the orbits of Mars and Jupiter (see Figure 8-13 on page 181, again). Some scientists think that the asteroids are materials left over from the birth of our solar system that never combined to form a planet.

A *comet* is a loose mass of rock, ice, dust, and gases that moves through space as a unit. Comets travel in stretched-out orbits (see Figure 8-19). As a comet approaches the sun, energy from the sun makes the comet glow and produce a "tail." The comet's tail always points away from the sun, regardless of the direction in which the comet is moving.

A *meteoroid* is a fragment of rock that travels through space. Sometimes these fragments enter Earth's atmosphere at high speed. Contact with the atmosphere creates friction that causes the meteoroid to heat up and burn. This produces a bright streak across the night sky, called a **meteor** or

Figure 8-19. *Typical orbit of a comet.*

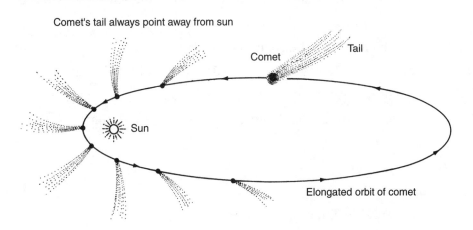

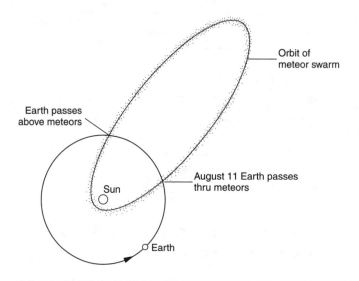

Figure 8-20 labels:
- Orbit of meteor swarm
- Earth passes above meteors
- August 11 Earth passes thru meteors
- Sun
- Earth

Figure 8-20. *When Earth passes through a meteor swarm orbiting the sun, a meteor shower occurs. If Earth passes above or below the swarm, no meteor shower occurs.*

a "shooting star." Occasionally, a large meteoroid does not burn up completely and a chunk of rock, called a *meteorite*, reaches Earth's surface.

A single orbit around the sun can contain many meteoroids. If Earth's orbit crosses the path of the meteoroids each year, a predictable meteor shower will occur (see Figure 8-20). On these nights, many meteors will occur in a specific area of the night sky. The Perseid meteor shower occurs about August 11 when Earth passes through the orbit of a large swarm of meteoroids.

 ## Review Questions

Multiple Choice

16. Carol saw a "shooting star" in the night sky. What she actually saw was a
    (1) comet (2) meteor (3) planet (4) moving star

17. The major members of the solar system are
    (1) comets and meteoroids   (3) the sun and the moon
    (2) asteroids and satellites   (4) the sun and the planets

18. Earth is apparently the only planet in our solar system that
    (1) has a moon
    (2) revolves around the sun
    (3) supports life
    (4) has a day side and a night side

19. Cheryl wanted to photograph a full moon on January 1, but the night was cloudy. When will be her next opportunity to take a picture of a full moon?

(1) January 15   (2) January 30   (3) February 10   (4) February 18

20. The moon shines by

(1) its own light

(3) light reflected off Earth

(2) radioactivity

(4) reflecting sunlight

21. Diagram (a) shows how the moon looked on three nights in May:

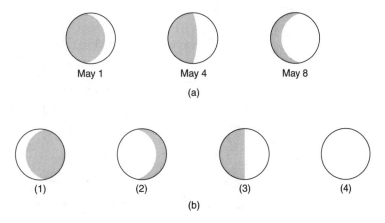

Which figure in diagram (b) shows how the moon would most likely appear on the night of May 15?

(1)   (2)   (3)   (4)

22. As the moon revolves around Earth, the moon

(1) always keeps the same side facing Earth

(2) does not rotate on its axis

(3) turns its entire surface toward Earth

(4) does not reflect sunlight

23. At which position in the moon's orbit can all of the moon's lighted side be seen from Earth?

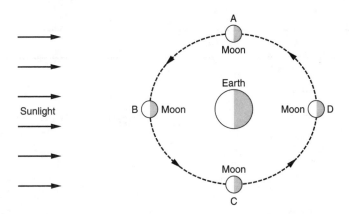

(1) A   (2) B   (3) C   (4) D

24. A solar eclipse can take place only when the moon is in

(1) full moon phase   (3) new moon phase

(2) last quarter phase   (4) first quarter phase

25. The diagram below suggests that, in causing tides on Earth, the moon's gravity

(1) has more effect than the sun's gravity

(2) has less effect than the sun's gravity

(3) is equal to the sun's gravity

(4) has no significant effect

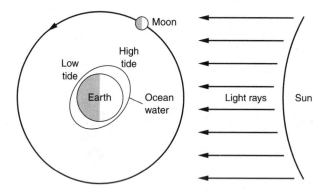

26. The most extreme high and low tides occur when

(1) the sun, Earth, and moon are at right angles to one another

(2) there has been a heavy rainstorm

(3) it is nighttime

(4) the sun, Earth, and moon are lined up

Thinking and Analyzing

27. The table below shows the times at which high and low tides occurred over two days. Based on the pattern in the table, at about what time will the next low tide occur?

Tidal Data Table

Day	Time	Tide
1	4:20 A.M.	Low
	10:35 A.M.	High
	4:45 P.M	Low
	11:00 P.M	High
2	5:10 A.M.	Low
	11:25 A.M.	High
	5:35 P.M.	Low
	11:50 P.M.	High

28. The diagram shows the four inner planets at various positions in their orbits. (a) Why is Mars visible from Earth in the night sky? (b) Why is Mercury not visible from Earth?

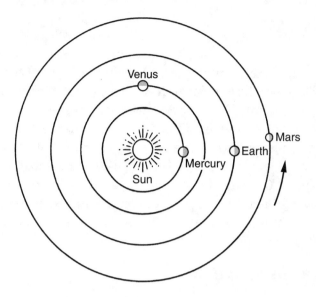

(c) When is Venus visible from Earth?

(1) before sunrise   (2) before sunset   (3) noon   (4) midnight

29. The table below gives the length of one rotation for the first five planets. (a) Which planet is spinning most rapidly on its axis? (b) One day on Earth is almost equal in length of time as one day on what other planet?

Planet	Length of One Rotation
Mercury	59 days
Venus	243 days
Earth	24 hours
Mars	24.6 hours
Jupiter	9.9 hours

30. Jesse saw a group of starlike objects in the night sky on April 7. A month later he noticed that one of the objects had moved to a different position (see the diagram). What was the object that moved most likely to be?

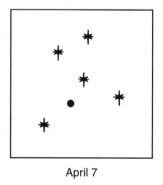

April 7

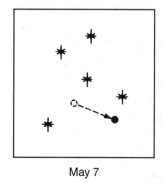

May 7

Our Place in the Universe

The Universe

The *universe* consists of space and all matter and energy scattered throughout space. There are far more stars in the universe than there are grains of sand on all the beaches on Earth. Nevertheless, the distances between stars are so vast that most of the universe is empty space.

Stars are not evenly distributed in space, but are clustered together in large groups called *galaxies*. The universe contains billions of galaxies, and each galaxy contains billions of stars. Galaxies are separated by great distances in space. Our sun is a star in the galaxy called the *Milky Way*. From Earth on a clear night, we can see only a few thousand nearby stars in our own galaxy. Astronomers must use telescopes to see stars in other galaxies.

Distances in Space

Distances in space are so great that they are difficult to fully understand. For example, the distance from Earth to our nearest neighbor star (besides the sun) is about 41,000,000,000,000 (41 trillion) kilometers! Because such numbers are so large, astronomers use a unit called a light-year to express distances to the stars. A **light-year** is the distance light travels in one year, about 9.5 trillion kilometers (see Figure 8-21). Using light-years allows us to describe distance to stars with more convenient size numbers. Our sun, the closest star to Earth, is a little more than 8 light-minutes from Earth. Our nearest neighbor star is 4.3 light-years away. Many other stars are billions of light-years away. This means we are just seeing the light that left these stars more than a billion years ago.

Figure 8-21. *A beam of light travels about 9.5 trillion kilometers per year. The light from a star that is 47.5 trillion kilometers from Earth takes 5 years to reach us.*

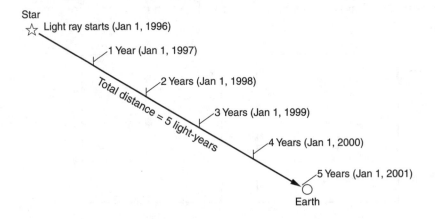

Many stars are as large or larger than our sun. However, because they are so far away, they appear as points of light in the night sky.

Formation of the Solar System

Although astronomers have some evidence to support their theory of the origin of the universe and our solar system, many of the details are still missing. Scientists think that between 15 and 20 billion years ago, a violent explosion created the universe, producing all the material from which the stars, planets, satellites, and other objects formed. This idea is called the *big bang theory*. If such an explosion did occur, one would expect to find all the galaxies moving outward and away from one another; this is exactly what astronomers have found.

Scientists further think that the expanding gases and dust particles condensed into local systems to form galaxies filled with stars. About 5 billion years ago, a dense area in the center of our local cloud of rotating gases and dust became our star, the sun (see Figure 8-22). The sun contains about 99 percent of the original cloud material. Soon afterward, the remaining material condensed into the planets, satellites, comets, meteoroids, and asteroids. The age of the oldest rocks from Earth, the moon, and meteorites is 4.5 billion years for all of them. This is evidence of a similar time of origin.

Figure 8-22. *(a) Cloud of gas and dust in space, (b) Condensation of cloud, (c) Early sun and planet formation, (d) Present-day solar system.*

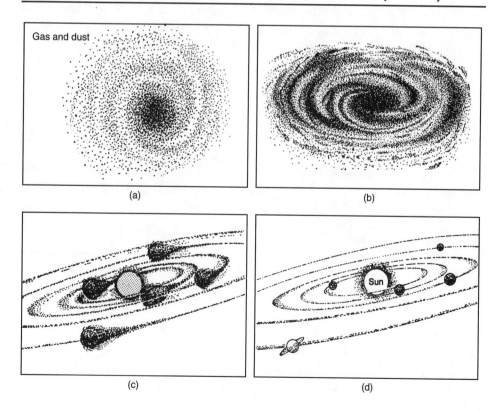

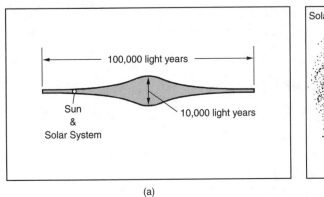
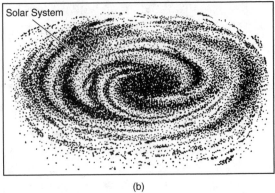

| (a) | (b) |

Figure 8-23. *The Milky Way galaxy and the location of the solar system.*

Earth's Position in the Universe

Our sun is one of 800 billion stars located in a star cluster called the Milky Way Galaxy. The Milky Way Galaxy is a large, disk-shaped group of stars that somewhat resembles a pinwheel. It is 100,000 light-years across and about 10,000 light-years thick in the center. As shown in Figure 8-23, our sun, along with Earth, all the other planets, satellites, asteroids, and comets, is located about two-thirds out in one of the spiral arms of the galaxy.

On a clear night we can see a few thousand neighbor stars all within our galaxy. If we observe carefully, we can see a band of closely packed stars that appear as a haze, or cloud across the night sky. This milky-like patch across the sky contains many stars also in the Milky Way Galaxy. Despite the fact that stars appear near each other, the distances between them are vast. Other galaxies and their stars are so distant that a telescope is necessary to see them. When you look at the night sky, you are seeing stars in a very small portion of the universe that is near us.

Stars

Table 8-5 lists the general characteristics of stars. Astronomers think that stars begin as clouds of gas and dust in space. Gravity causes the ma-

Table 8-5. *General Star Characteristics*

Distance:	Sun	150,000,000 km
	Neighbor star	4.3 light-years
	Distant star	Millions of light-years
Composition		Hydrogen and helium
Size		100 km–several billion km
Temperature (surface)		3000°-30,000°C

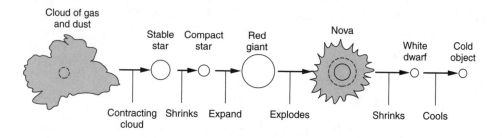

Figure 8-24. *Life cycle of a typical star.*

terial to contract and form the nucleus of a new star that begins radiating energy. Over millions of years, the new star becomes hot enough to glow. Stars, like planets, are nearly spherical.

In this early developmental stage, stars appear reddish because they are relatively cool. Depending on the amount of material present, the star will continue to grow in size and temperature. As it becomes hotter, its surface color changes from red, to orange, to yellow, and possibly to white. If the star is massive enough, it will become bright blue-white. Eventually, the cloud of gas and dust is used up and the star stabilizes. Our sun is an average-size, yellow star in this stable stage of development (see Figure 8-24).

The thermonuclear reaction that changes hydrogen into helium continues to release energy in the form of light and heat. When the star's hydrogen is exhausted, the star collapses. Collapsing compresses the star and causes it to heat and rapidly expand into a cooler red giant or supergiant.

The giant star eventually explodes, losing its outer layer with a burst of energy. This bright flare-up of a star is called a *nova*. Finally, the star will shrink into a white dwarf and evolve into a cold object.

The amount of time it takes for a star to go through its life cycle depends on the original amount of material in the cloud and the size of the star. Our sun's life span is estimated to be about 10 billion years. Our sun appears to have another 5 billion years of life left.

 Review Questions

Multiple Choice

31. Evidence for the big bang theory includes the fact that

    (1) the universe is shrinking

    (2) all galaxies are moving away from Earth

    (3) stars and galaxies are moving toward Earth

    (4) stars are exploding

32. Analyses of meteorites, Earth rocks, and moon rocks indicate a maximum age of 4.5 billion years. This is evidence that all three bodies probably formed

(1) independently

(2) as a single body in the solar system

(3) less then 4.5 billion years ago

(4) at the same time

33. The Milky Way appears as a hazy band of light that stretches across the night sky. Actually, it is our view outward of our galaxy. The hazy band of light is produced by

(1) sunlight

(2) reflected moonlight

(3) stars in our galaxy

(4) distant city lights

34. The sun and our solar system are located

(1) in the center of the Milky Way Galaxy

(2) within the spiral arms of the Milky Way Galaxy

(3) outside of the Milky Way Galaxy

(4) on the outer edge of the Milky Way Galaxy

35. Stars appear to have life spans of billions of years. During that time they

(1) never change

(2) constantly get larger

(3) constantly get smaller

(4) change in size and temperature

36. Our sun is a stable, yellow star. It is considered to be a(an)

(1) young star

(2) middle-aged star

(3) old star

(4) nova star

Review Star Data Table and answer questions 37 and 38.

Star Data Table

Star	Absolute Magnitude (Brightness)	Apparent Magnitude (Brightness)	Distance
Sirius	1.4	−1.5	8.8
Vega	0.5	0.04	26.4
Capella	−0.6	0.05	45.6
Antares	−4.7	1.0	423.8

37. What units of measurement do the numbers in the Distance column most likely represent?

(1) miles   (2) kilometers   (3) light-years   (4) years

38. Apparent magnitude is how bright a star appears when viewed from Earth. The lower the magnitude number the brighter the star. Which star appears the brightest when viewed from Earth?

(1) Sirius (2) Vega (3) Capella (4) Antares

39. Billions of stars clustered together in large groups are called

(1) solar systems (2) galaxies (3) asteroid belts (4) constellations

40. Which of the following objects is the largest?

(1) sun (2) galaxy (3) solar system (4) Earth

Thinking and Analyzing

Review the *Distances of Some Objects from Earth* table and answer questions 41 and 42.

Distances of Some Objects from Earth

Object	Average Distance
Sun	149,600,000 km
Moon	384,000 km
Pluto's Orbit	5,750,000,000 km
Neighbor Star	4.3 light-years
Neighbor Galaxy	2,200,000 light-years
Distant Galaxy	65,000,000 light-years

41. How do the distances between the planets in our solar system compare with the distance between stars?

42. Why do astronomers use light-years to measure distance to stars and galaxies?

Review the *Star Classification* table and answer question 43.

Star Classification

Class	Color	Surface Temperature	Example
O	blue	30,000°C–60,000°C	No common star
B	blue-white	10,000°C–30,000°C	Rigel
A	white	7500°C–10,000°C	Sirius
F	white-yellow	6000°C–7500°C	Polaris
G	yellow	5000°C–6000°C	Sun
K	orange	3500°C–5000°C	Arcturus
M	red	2400°C–3500°C	Betelgeuse

43. To what is the color of a star most closely related?

Chapter 9
Matter

Points to Remember

▷ Matter is anything that has mass and occupies space. Matter can exist as a solid, liquid, or gas.

▷ Elements are substances that cannot be broken down into simple substances. Two or more elements can combine to form a compound.

▷ Atoms are composed of protons, neutrons, and electrons.

▷ Changes in matter may be physical or chemical. During a chemical change, new substances are formed. Mixtures do not form new substances. Solutions are mixtures. The parts of a mixture can be separated by physical changes such as filtering and evaporating.

▷ Density $= \dfrac{\text{mass}}{\text{volume}}$. Floating and sinking depend on differences in density.

▷ Matter can neither be created nor destroyed in a chemical reaction. Energy can either be absorbed or released during a chemical change. The rates of chemical reactions and dissolving are influenced by such factors as temperature and particle size.

Defining Matter

Look around you. The objects you see, such as this book, your desk and chair, and the walls and ceiling, are all made of matter. The air (a mixture of gases) that surrounds you is also made of matter. In fact, every solid, liquid, and gas is a form of matter.

Matter is defined as anything that has mass and takes up space. *Mass* is the total amount of material in an object. We measure mass with a triple-beam balance or an equal-arm balance, as shown in Figure 9-1. Notice that a balloon filled with air has a greater mass than an empty balloon, because air has mass. The amount of space an object occupies is called its *volume*. The air in the filled balloon in Figure 9-1 takes up space, giving the balloon a greater volume than the empty balloon.

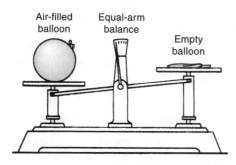

Figure 9-1. *The air-filled balloon is heavier and takes up more space than the empty balloon because air is matter.*

Is there anything that is not made of matter? Is there anything that has no mass and takes up no space? Figure 9-2 shows that shining a light on a balance has no effect on the balance. This is because light is a form of energy. Energy is not matter, since it has no mass and no volume. Some other forms of energy are heat and sound.

Figure 9-2. *The balance is unaffected by the light shining on it because light is not matter.*

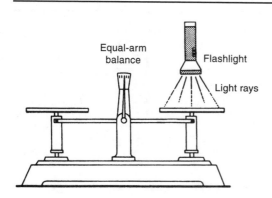

Elements

The basic building blocks of matter are called **elements.** All substances are made up of one or more elements. Oxygen, hydrogen, gold, and iron are examples of elements. Each element is represented by a symbol made up of one or two letters. For example, the symbol for hydrogen is H, oxygen is O, and gold is Au. The first letter of the symbol is always capitalized while the second letter, if any, is always lower case. There are at least 115 known elements. (Note: The number of elements changes as new elements are created in research laboratories.) However, fewer than half of them occur commonly in nature. Table 9-1 lists the most common elements found in Earth's crust.

The smallest particle of an element that has the properties of that element is called an **atom.** All atoms of a particular element are alike, but they are different from the atoms of any other element. For instance, all hydrogen atoms are alike, but they differ from oxygen atoms. If there are 115 different elements, there are 115 different kinds of atoms.

Atomic Structure

All atoms are composed of smaller *subatomic* particles. These particles are called *protons, neutrons,* and *electrons.* Protons, neutrons, and electrons differ in their mass, electrical charge, and location in the atom. Protons and neutrons have roughly the same mass, while electrons are much lighter. Protons have a positive (+) charge, and electrons have a negative (−) charge. Neutrons have no electrical charge: they are electrically neutral.

Protons and neutrons are found in the center, or **nucleus**, of the atom. Electrons orbit the nucleus, moving very rapidly. The negatively charged electrons are attracted to the positively charged protons in the nucleus because oppositely charged particles attract each other. Since like charges repel and all electrons are negatively charged, electrons repel other electrons. Table 9-2 summarizes the properties of the subatomic particles.

The atoms of different elements have a different number of protons in their nucleus. Oxygen has 8 protons, while carbon has 6, and uranium has 92. The number of protons in the nucleus is called the *atomic number.*

Table 9-1. *Most Common Elements in Earth's Crust*

Element	Chemical Symbol
Oxygen	O
Silicon	Si
Aluminum	Al
Iron	Fe
Calcium	Ca
Sodium	Na
Potassium	K
Magnesium	Mg

Table 9-2. *Properties of the Subatomic Particles*

Particle	Mass (AMU)*	Charge	Location
Proton	1	+	Nucleus
Neutron	1	0	Nucleus
Electron	0.00054 (often rounded to 0)	−	Outside the nucleus

*The Atomic Mass Unit (AMU) is a special unit created for measuring the mass of very small particles.

The Periodic Table of the Elements arranges the elements according to their atomic number.

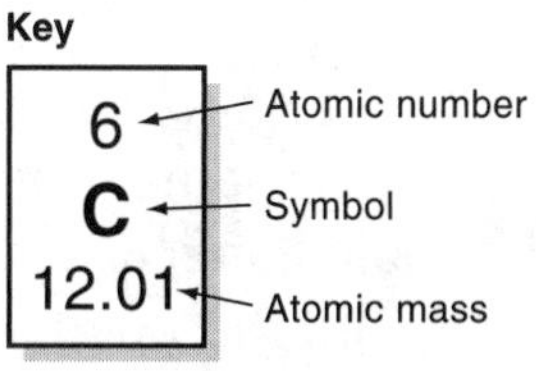

The Periodic Table of the Elements

Scientists organize the elements based on their properties on a chart known as *The Periodic Table of the Elements* (see Figure 9-3). On this table, elements with similar properties are placed in the same vertical column called a group. These groups are numbered from 1 through 18.

Figure 9-3. *The Periodic Table of the Elements.*

The majority of elements are shiny solids that conduct electricity. These elements are called *metals*. (Mercury, which is a liquid at room temperature, is also considered a metal.) A smaller number of elements are poor conductors of electricity and lack the luster of the metals. These are called *nonmetals*. A still smaller group of elements, which are all gases at room temperature, seldom react with other elements. These are called the *noble gases.*

The periodic table contains a zigzag line that separates the metals to the left from the nonmetals to the right. The last group of elements, group 18, contains the noble gases.

Process Skill

Reading for Understanding

The Tooth Fairy Project

In a study called "The Tooth Fairy Project" scientists have asked parents to send them their children's baby teeth. These teeth are being tested for a radioactive form of the element strontium (symbol: Sr) called strontium-90.

Radioactive substances, such as strontium-90, are extremely dangerous when absorbed into the body. Scientists have linked exposure to radioactivity to an increased number of cases of cancer.

Strontium-90 is an especially dangerous element because it is chemically similar to the element calcium. (Strontium is in the same group of elements as the element calcium—Group 2.) The body is fooled, mistakes the strontium-90 for calcium, and deposits the radioactive element into the bones and teeth.

By measuring the amount of radioactive strontium in baby teeth, scientists are trying to determine our levels of radioactive exposure. They are looking to see if there is a link between areas of high radioactive exposure and areas with high cancer rates.

The Tooth Fairy Project has been going on for quite some time and has collected thousands of baby teeth for testing. The project will continue for many years to come.

QUESTIONS

1. What are scientists in the Tooth Fairy Project looking for in baby teeth?
   (1) calcium   (2) strontium   (3) cancer   (4) tooth decay
2. The Tooth Fairy Project is trying to link
   (1) radioactive exposure to cancer rates
   (2) strontium-90 to calcium
   (3) age to radioactivity
   (4) calcium to cancer rates
3. Referring to The Periodic Table of the Elements, explain why the body mistakes strontium for calcium.

Table 9-3. *Some Common Compounds and Their Chemical Formulas*

Compound	Formula	Elements
Table salt	NaCl	Sodium, Chlorine
Water	H_2O	Hydrogen, Oxygen
Sugar (sucrose)	$C_{12}H_{22}O_{11}$	Carbon, Hydrogen, Oxygen
Quartz	SiO_2	Silicon, Oxygen
Ammonia	NH_3	Nitrogen, Hydrogen

Compounds

Scientists know millions of different substances. How is this possible if there are only 115 elements? Elements can combine to form new substances. A substance that is formed when two or more different elements combine is called a **compound.** Since many different combinations of elements are possible, many different compounds can exist. The common substance water is a compound formed when the elements hydrogen and oxygen combine.

A compound is represented by a chemical formula that indicates which elements have combined, and in what proportions. The chemical formula for water, H_2O, indicates that water contains two atoms of hydrogen to every atom of oxygen. Table 9-3 lists some common compounds and their chemical formulas.

The smallest particle of a compound is called a **molecule.** A water molecule is composed of two hydrogen atoms and one oxygen atom, as shown in Figure 9-4. Atoms of the same element can also combine to form molecules. For example, two oxygen atoms combine to form a molecule of oxygen gas, O_2.

Atoms and molecules are extremely small. To get an idea of just how small, consider that 1 teaspoonful of water contains about 175 *sextillion* water molecules. (That would be written as 1.75×10^{23}, or 175 followed by 21 zeros!)

Figure 9-4. *The arrangement of atoms in a molecule of water.*

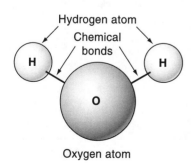

Chemical Bonds

Atoms in a molecule are joined together by a special link called a *chemical bond* (see Figure 9-4). These bonds store chemical energy. Sometimes this energy can be released by a chemical reaction. Burning is one type of chemical reaction that releases energy. When wood is burned, energy stored in the chemical bonds within the wood is released as heat and light. Respiration is another chemical reaction that releases energy from chemical bonds.

Mixtures

When two or more materials are put together and do not form a new substance, a mixture is formed. Saltwater, for example, is a mixture of salt and water. Sand is a mixture of minerals. Blood is a mixture of different cells, water, and other nutrients. Air is a mixture of several gases.

Unlike compounds, mixtures cannot be represented with a chemical formula. Table salt—a compound—is always NaCl. However, saltwater—a mixture—can be more or less salty and still be saltwater. For example, salt in Utah is exactly the same as salt in New York, but the saltwater in the Great Salt Lake in Utah is quite different from the saltwater in the Atlantic Ocean off New York.

Physical Properties

Have you ever mistaken salt for sugar? They look very much alike. How might you tell them apart? Scientists faced with similar problems identify substances by examining their *properties.*

A difference in taste helps you distinguish salt from sugar. A difference in color (as well as taste) helps you distinguish salt from pepper. Taste and color are **physical properties**—properties that can be determined without changing the identity of a substance. All substances have unique physical properties by which they can be identified. Table 9-4 lists some physical properties often used to identify substances.

Table 9-4. *Examples of Physical Properties of Substances*

Property	Example
Phase	Mercury is a liquid at room temperature.
Color	Sulfur is yellow.
Odor	Hydrogen sulfide smells like rotten eggs.
Density	Lead is much denser than aluminum.
Solubility	Salt dissolves in water.
Melting point	Ice melts at 0°C.
Boiling point	Water boils at 100°C.

Phases

One obvious physical property of a substance is whether it is a solid, a liquid, or a gas. These three forms of matter are called *phases*. The arrangement and motion of the molecules within a substance determine its phase.

1. In *solids*, the molecules are close together, move relatively slowly, and remain in fixed (unchanging) positions. A solid has a definite shape and a definite volume; that is, its shape and size do not depend on the container it is in.
2. In *liquids*, the molecules are usually farther apart and moving faster than the molecules in solids. The molecules in a liquid can change position and flow past each other. A liquid has no definite shape; it takes on the shape of its container. However, liquids do have a definite volume. A given quantity of a liquid takes up the same amount of space regardless of the shape and size of its container.
3. In *gases*, the molecules are much farther apart and move even faster than in liquids. The molecules of a gas can move anywhere within their container. A gas has no definite shape or volume but expands or contracts to fill whatever container it is in. Figure 9-5 shows how molecules are typically arranged in solids, liquids, and gases.

Changes in Phase

Since the phase of a substance depends on the arrangement of its molecules, a change in this arrangement can cause a change in phase.

1. To change a solid into a liquid, the molecules must generally be moved farther apart, out of their fixed positions. This is called **melting.** Heat energy must be added to a substance to separate its molecules, so energy is absorbed during melting. (See Table 9-5 on page 208.)
2. **Freezing** is the opposite of melting. When a liquid freezes into a solid, the molecules come together and bond more tightly into fixed positions. This process releases energy.

Figure 9-5. *The three phases of matter: solid, liquid, and gas.*

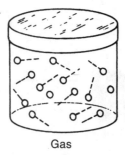

Solid   Liquid   Gas

Table 9-5. *Examples of Water Phase Changed*

Phase Change	Heat Flow	Examples
Liquid to Gas	Water absorbs heat energy	Puddle Evaporates, Boiling Water
Solid to Liquid	Water absorbs heat energy	Melting Snow, Ice Melting in Soda
Gas to Liquid	Heat energy is released	Cloud Condensation, Water on Kitchen Window in Winter
Liquid to Solid	Heat energy is released	Ice Cubes Form in Freezer, Ice Forming on Lake

3. Changing a liquid into a gas, by **boiling** or **evaporation**, requires that the molecules of the liquid be separated even farther. Energy is therefore absorbed when a liquid changes into a gas.

4. The change from a gas to a liquid is called **condensation.** During condensation, molecules of a gas move closer together to form a liquid, and energy is released. Figure 9-6 illustrates the energy changes associated with changes in phase.

For each substance, the change in phase from a solid to a liquid occurs at a particular temperature called its **melting point.** The melting point of ice (the solid form of water) is 0°C. The temperature at which a liquid freezes into a solid is called its freezing point. The **freezing point** of water is 0°C. The freezing point and melting point of a substance are always the same.

The temperature at which a liquid boils and changes rapidly into a gas is called its **boiling point.** The boiling point of water is 100°C. This is also the temperature at which water vapor, cooling from above 100°C, begins to condense into liquid water.

While a substance is changing phase, its temperature remains constant. For example, while you are boiling water, the temperature remains at 100°C even though you are constantly supplying heat. The heat is used to cause the phase change rather than a change in temperature.

Figure 9-6. *Energy changes occur during changes in phase.*

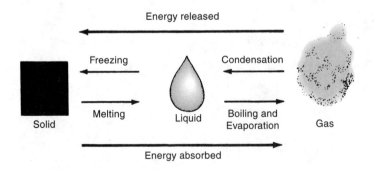

Process Skill

Interpreting Data in a Table

By using melting point and boiling point information, we can determine what phase a substance will be in at a given temperature. If the temperature of a substance is below its melting point, the substance is a solid. If the temperature is above its boiling point, the substance is a gas. If the temperature is between its melting and boiling points, the substance is a liquid. For example, at room temperature (20°C), water is a liquid, because 20°C is between the melting point and boiling point of water.

The table below lists the melting point and boiling point of some common substances. What phase would table salt be in at a temperature of 1000°C? Since 1000°C is above the melting point of salt but below its boiling point, table salt would be a liquid at that temperature. Use the same kind of reasoning, based on the table below, to answer the following questions.

Melting Points and Boiling Points of Some Common Substances

Substance	Melting Point (°C)	Boiling Point (°C)	Phase at 20°C (room temperature)
Water	0	100	Liquid
Alcohol	−117	78	Liquid
Table salt	801	1413	Solid
Oxygen	−218	−183	Gas

QUESTIONS

1. At a temperature of −190°C, oxygen is in the form of a
   (1) gas   (2) liquid   (3) solid
2. Alcohol would be a liquid at all of the following temperatures except:
   (1) −100°C   (2) 32°C   (3) 100°C   (4) 77°C
3. The only substance listed that could be a liquid at a temperature of 90°C is
   (1) table salt   (2) water   (3) alcohol   (4) oxygen

Review Questions

Multiple Choice

1. What does the diagram below show about matter?

   (1) Matter is made up of elements.

   (2) Matter takes up space.

   (3) Matter is a solid.

   (4) Matter has mass.

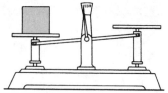

2. Which is not an example of matter?

    (1) water   (2) air   (3) gold   (4) sound

3. The amount of space an object occupies is called its

    (1) volume   (3) weight
    (2) mass   (4) length

4. Atoms in a molecule are joined together by

    (1) chemical bonds   (3) electricity
    (2) magnetism   (4) gravity

5. The circles in the closed jars shown below represent particles of matter. Which jar most likely contains a solid?

(1)   (2)   (3)   (4)

6. In which phase of matter are the particles farthest apart and moving the fastest?

    (1) solid   (2) liquid   (3) gas

7. When you pour water from a beaker into a flask, there is a change in its

    (1) mass   (2) volume   (3) shape   (4) density

8. Why are atoms electrically neutral?

    (1) Subatomic particles have no charge.

    (2) The number of negative electrons equals the number of positive protons.

    (3) Atoms contain only neutrons, which have no charge.

    (4) Protons are found in the nucleus, while electrons are found outside the nucleus.

9. Condensation refers to a phase change from

    (1) solid to liquid   (3) liquid to gas
    (2) liquid to solid   (4) gas to liquid

10. The temperature at which a substance melts is the same temperature at which it

    (1) boils   (2) freezes   (3) condenses   (4) evaporates

Thinking and Analyzing

11. In two or more sentences, explain the differences between a mixture and a compound. Give two examples of each.

12. A student opens a bottle of perfume in the back of the classroom. Although you do not see the bottle being opened, you become aware of the scent. In one or two sentences, tell how this effect happens.

Refer to *The Periodic Table of the Elements* on page 203 to answer questions 13–14.

13. Which of the following elements can be classified as a metal?
    (1) O (oxygen)   (3) Na (sodium)
    (2) He (helium)   (4) Cl (chlorine)

14. Which element is a poor conductor of electricity?
    (1) Fe (iron)   (2) Cu (copper)   (3) Au (gold)   (4) S (sulfur)

15. Give an example of a phase change that releases energy.

16. The element sodium is very similar to the element potassium and very different from the element chlorine. Using The Periodic Table of the Elements as a guide, explain why.

Solutions

How does a mixture of salt and water differ from a mixture of sand and water? When salt is placed in water, the particles of salt disappear, yet they can still be detected when tasting the mixture. A **solution** is a mixture in which the components remain evenly distributed. Saltwater is a solution. In a mixture of sand and water, the sand remains clearly visible settled at the bottom of the container. Sand and water is not a solution. The physical property that distinguishes the sand from the salt is called **solubility**. We say that salt dissolves in water to form a solution while sand does not dissolve in water. Salt is **soluble** in water while sand is **insoluble** in water.

A solution generally has two parts, the **solute** and the **solvent**. The solvent does the dissolving while the solute gets dissolved. For example, when a solid dissolves in a liquid, the solid is the solute and the liquid is the solvent. Gases or other liquids may also dissolve in liquids to form a solution. Table 9-6 shows some common solutions with water as the solvent.

Table 9-6. *Some Solutes That Dissolve in Water.*

Solution	Solute	Phase of Solute
Seltzer	Carbon dioxide	Gas
Tea	Tea	Solid
Vodka	Alcohol	Liquid

Rate of Dissolving

What do you do after you add sugar to a cup of tea? You probably stir the mixture. Why? Stirring is one method of increasing the rate of dissolving. Which would dissolve faster, a sugar cube or a packet of granulated sugar? The smaller the particle size is, the faster the dissolving process will be. Therefore, granulated sugar dissolves faster than a lump of sugar. Even granulated sugar dissolves quite slowly in iced tea. It dissolves much faster in hot tea. An increase in temperature usually increases the rate of dissolving.

Solubility

Not only does an increase in temperature dissolve sugar faster, it also allows more sugar to dissolve. In fact, it is possible to dissolve two cups of sugar in one cup of water if the water is hot enough (100°C).

In general, raising the temperature increases the amount of solid solute that can dissolve in a liquid. The maximum amount of solute that can dissolve in a given amount of solvent is called the *solubility*. Generally, the solubility of a solid in a liquid increases as temperature increases. Gases however, behave differently. The solubility of a gas in a liquid decreases when temperature increases.

The solubility of a gas is also affected by pressure; the higher the pressure, the more soluble the gas. When you open a bottle of soda, you decrease the pressure on the solution. The carbon dioxide gas that was dissolved at the higher pressure comes out of the solution, forming bubbles.

Choosing a Solvent

You have probably heard the expression, "Oil and water don't mix." A chemist might say instead, "Oil is not soluble in water." Oil is soluble in other solvents. We often need to choose a suitable solvent for a given solute. For example, nail polish does not dissolve in water. It does dissolve in acetone, a liquid often used as a nail polish remover. Grease and oil, which often stain clothing, do not dissolve in water. The "dry cleaners" use a liquid called "perc" that dissolves the grease without harming the fabric. The solubility of a given solute in a given solvent depends upon the chemical bonds in the two substances.

Density

Why are airplanes made of aluminum, and fishing sinkers made of lead? You might answer that aluminum is a light metal, while lead is a heavy metal. Yet an aluminum airplane has a much larger mass than a lead fishing sinker. When we say that lead is heavier than aluminum, we really mean that when these two pieces of metal are the same size, the lead piece will be heavier. (If the pieces are not the same size, we need another way to compare them.) The quantity that compares the mass of an object to its size, or more specifically its volume, is called *density*.

Density is defined as the mass of an object divided by its volume $\left(\text{density} = \dfrac{\text{mass}}{\text{volume}}\right)$. While the mass and volume of a piece of metal depend on the size of the piece, the density depends on only the nature of the metal and its temperature. (We will discuss the effect of temperature later on.) Let's compare the densities of lead and aluminum. At room temperature, the density of aluminum is 2.7 grams per cubic centimeter, which we abbreviate as 2.7 g/cm³. Since density is mass divided by volume, the unit of density contains a mass unit, grams, divided by a volume unit, cubic centimeters. (Another unit of volume the milliliter, abbreviated mL, is often used when measuring the volume of a liquid.) When we say that aluminum has a density of 2.7 g/cm³, this means that a piece of aluminum with a volume of 1 cubic centimeter has a mass of 2.7 grams. The density of lead at room temperature is 11.3 g/cm³. Lead is about four times as dense as aluminum.

Which is heavier, 10 grams of lead or 10 grams of aluminum? Of course, this is a trick question. Since both are 10 grams, they are equally heavy. Which has a larger volume, 10 grams of lead or 10 grams of aluminum? This question requires some thinking. The density of aluminum is 2.7 g/cm³, so 10 grams of aluminum has a volume greater than 1 cubic centimeter. Table 9-7 shows how the volume may be calculated. The density of lead is 11.3 g/cm³, so a 10-gram piece of lead has a volume of less than 1 cubic centimeter. In comparing objects of equal mass, the denser object is the one with the

Table 9-7. *Performing Calculations with Density*

Finding mass from volume and density	$\text{Density} = \dfrac{\text{Mass}}{\text{Volume}}$
To get mass by itself, multiply both sides of the equation by volume.	$\text{Volume} \times \text{Density} = \dfrac{\text{Mass}}{\text{Volume}} \times \text{Volume}$
Volume cancels on the right side, so we get	$\text{Mass} = \text{Density} \times \text{Volume}$
Finding volume from mass and density	$\text{Density} = \dfrac{\text{Mass}}{\text{Volume}}$
To get volume out of the denominator, multiply both sides by volume.	$\text{Volume} \times \text{Density} = \dfrac{\text{Mass}}{\text{Volume}} \times \text{Volume}$
Volume cancels on the right side, so;	$\text{Volume} \times \text{Density} = \text{Mass}$
To get volume by itself, divide both sides by density.	$\text{Volume} \times \dfrac{\text{Density}}{\text{Density}} = \dfrac{\text{Mass}}{\text{Density}}$
Density cancels on the left side, so we get:	$\text{Volume} = \dfrac{\text{Mass}}{\text{Density}}$

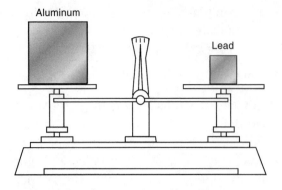

Figure 9-7. *These metal cubes are of equal mass; the cube with the smaller volume has the greater density.*

smaller volume. Figure 9-7 compares pieces of lead and aluminum of the same mass.

Which is heavier, 10 cm³ of aluminum, or 10 cm³ of lead? When comparing objects of equal volume, the object with the greater density has the greater mass. Thus, 10 cm³ of lead weighs about four times as much as the same volume of aluminum, since lead is about four times denser than aluminum. Figure 9-8 compares pieces of lead and aluminum that have the same volume.

The density of a 10-cm³ piece of lead is 11.3 g/cm³. What is the density of a 20-cm³ piece of lead? The answer: still 11.3 g/cm³! Table 9-8 gives the mass, volume, and density of several pieces of lead. As you can see, the density remains the same no matter what size the piece of metal is. The density of a material does not depend on the size of the object. This makes density a very useful property for identifying materials.

Using Density to Identify a Metal

Tinfoil and aluminum foil look very much alike. Aluminum has a density of 2.7 g/cm³, while tin has a density of 7.3 g/cm³. A chemist measures the vol-

Figure 9-8. *These metal cubes are of equal volume; the cube with the greater density has the greater mass.*

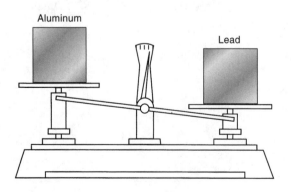

Table 9-8. *Mass, Volume, and Densities of Several Pieces of Lead*

Mass	Volume	Density (Mass/Volume)
113 g	10 cm³	11.3 g/cm³
226 g	20 cm³	11.3 g/cm³
1130 g	100 cm³	11.3 g/cm³

ume of a sample of metal foil and finds it to be 5.0 cm³. He then weighs it and finds its mass to be 36.5 g. Is it aluminum or tin? Remember that density is mass/volume. To find the density of the metal, divide the mass, 36.5 grams, by the volume, 5.0 cm³. Thus,

$$\text{density} = \frac{36.5 \text{ g}}{5.0 \text{ cm}^3} = 7.3 \text{ g/cm}^3.$$

The metal was tin.

Why Do Objects Float?

A wooden log floats on water, while an iron nail sinks. Why? The answer to this question lies in the densities of these materials. A material will float if it is less dense than the liquid in which it is placed. From this information, we can conclude that iron is more dense than water, while wood is less dense than water. Water has a density of 1 g/cm³. Any object with a density greater than 1 g/cm³ will sink in water. The density of iron is 7.9 g/cm³, and therefore it sinks in water.

Some liquids do not mix with each other. For example, the oil and vinegar in salad dressing form separate layers, as shown in Figure 9-9. What can you conclude about the density of oil compared with the density of vinegar? Since the oil floats over the vinegar, the oil must be less dense.

Helium balloons float on air in much the same way that wood floats on water. The density of a helium balloon is much less than the density of air.

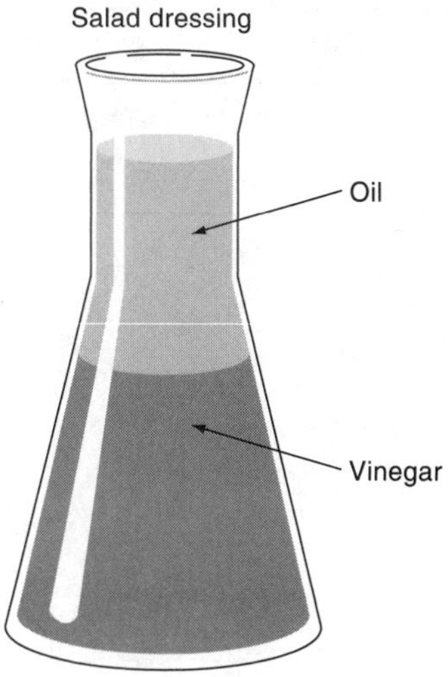

Salad dressing

Oil

Vinegar

Figure 9-9. *The less dense liquid (oil) floats on top of the more dense liquid (vinegar).*

Some balloons use hot air instead of helium. Since hot-air balloons float, hot air must be less dense than cold air. When most materials are heated, they expand. This means that their volume increases while their mass stays the same. Since density is mass/volume, an increase in volume will cause a decrease in a material's density. In general, an increase in temperature causes a decrease in density.

Objects that float are said to be **buoyant**. The **buoyancy** of an object in water depends on the density of the object relative to the density of the water. The less dense the object is and the denser the fluid is, the greater the buoyancy of the object.

 Review Questions

Multiple Choice

17. Which one of the following materials is soluble in water?

(1) oil   (2) salt   (3) sand   (4) grease

18. Which of the following would dissolve the fastest?

(1) sugar cubes in iced tea   (3) sugar cubes in hot tea

(2) granulated sugar in iced tea   (4) granulated sugar in hot tea

19. When dissolving salt in water, the salt is considered the

(1) solvent   (2) solute   (3) mixture   (4) solution

Thinking and Analyzing

Use the information below to answer questions 20–25.

A student measures the masses and volumes of four pieces of metal. The results are shown below.

Property	Metal A	Metal B	Metal C	Metal D
Mass	10.0 g	10.0 g	30.0 g	40.0 g
Volume	2.0 cm³	5.0 cm³	5.0 cm³	8.0 cm³

20. Which metal is the most dense?

(1) Metal A   (2) Metal B   (3) Metal C   (4) Metal D

21. Which two pieces might be made of the same metal?

(1) A and B   (2) B and C   (3) A and D   (4) C and D

22. If we compared 10.0-gram samples of all four metals, which would have the largest volume?

(1) Metal A   (2) Metal B   (3) Metal C   (4) Metal D

23. If we compared samples of the metals with the same volumes, which would have the largest mass?

    (1) Metal A   (2) Metal B   (3) Metal C   (4) Metal D

24. The density of a 100.0-gram sample of metal A should be

    (1) 5.0 g/cm³   (2) 10 g/cm³   (3) 50 g/cm³   (4) 100 g/cm³

Questions 25 and 26 refer to the diagram below, which represents a beaker of ice water.

25. This diagram indicates that the ice is

    (1) less dense than water   (3) colder than water

    (2) more dense than water   (4) warmer than water

26. As water freezes to ice, its

    (1) volume decreases   (3) volume increases

    (2) mass decreases   (4) mass increases

Questions 27 and 28 refer to the diagram below, which shows the relative densities of some liquids and solids.

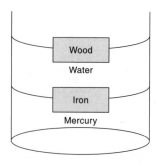

27. Which liquid in the diagram is the least dense?

    (1) wood   (2) water   (3) iron   (4) mercury

28. Which solid in the diagram is the most dense?

    (1) wood   (2) water   (3) iron   (4) mercury

29. What effect does an increase in temperature have on the density of air? Explain your answer.

Changes in Matter

We see change around us all the time. There are changes in us. There are changes in the weather. There are changes on Earth. Some of these changes involve physical aspects of matter. Other changes involve chemical aspects.

Physical Changes

As you know, the chemical formula for water is H_2O because each water molecule is made up of two atoms of hydrogen and one atom of oxygen. What is the formula for ice? When water freezes, the arrangement of its molecules changes, but the molecules themselves do not change. They are still H_2O. A change of phase, such as freezing or melting, does not produce any new substances. A change that does not result in the formation of any new substances is a ***physical change.*** All changes of phase are physical changes. Crushing ice cubes into small pieces is also a physical change, since both crushed ice and ice cubes are made of the same substance.

Similarly, when you dissolve sugar in water, the sugar still tastes sweet and the water is still wet. No new substances have been formed, so dissolving is a physical change. Figure 9-10 shows why boiling, melting, and dissolving are physical changes.

Figure 9-10. During physical changes, no new substances are formed.

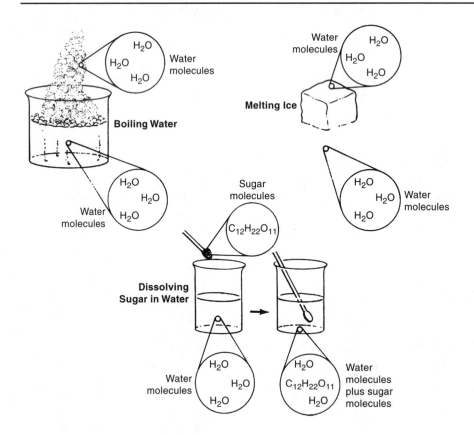

Separating the Parts of a Mixture

A mixture of two or more substances can be separated through physical changes. For example, a mixture of salt and water can be separated by evaporating the water, which leaves the salt behind. A mixture of iron and silver can be separated with a magnet. The iron will be attracted to the magnet while the silver will not. How might a mixture of salt and sand be separated? This would involve a series of physical changes. First, the mixture could be added to water. The salt would dissolve while the sand would not. This mixture could then be filtered, separating the undissolved sand from the saltwater solution. The dissolved salt particles are too small to be trapped on the filter paper, and pass through the tiny openings in the paper. The sand particles are too large to pass through the openings in the filter paper. Finally, we can boil off the water from the saltwater mixture leaving the salt behind. The sequence of steps is illustrated in Figure 9-11.

Chemical Changes

What happens if you forget to put a carton of milk back into the refrigerator? First, the milk gets warm. This is a physical change. However, if you leave the milk out too long, it turns sour. The sour taste is caused by the production of a new substance called *lactic acid.* A change that produces one or more new substances is called a ***chemical change***. When a chemical change occurs, we say there was a chemical reaction. Burning paper produces smoke and ash, both of which are new products. Burning is always a chemical change.

Figure 9-11. *Separating the parts of a mixture: (A) A mixture of salt and sand. (B) Dissolving the salt in water. (C) Separating the sand with a filter. (D) Evaporating the water.*

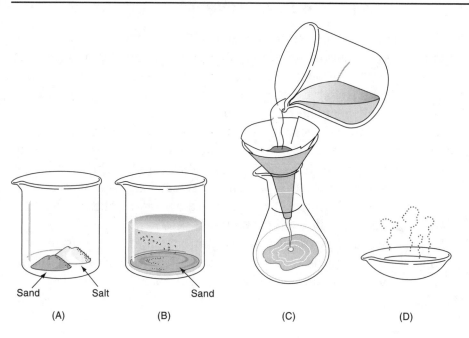

Sand   Salt   Sand

(A)   (B)   (C)   (D)

Table 9-9. *Examples of Chemical Changes*

Chemical Change	Reactants	Products	Equation
Burning coal	Carbon (C) + oxygen gas (O_2)	Carbon dioxide gas (CO_2)	$C + O_2 \rightarrow CO_2$
Rusting of iron	Iron (Fe) + oxygen gas (O_2)	Rust (Fe_2O_3)	$4Fe + 3O_2 \rightarrow 2Fe_2O_3$
Tarnishing of silver	Silver (Ag) + sulfur (S)	Tarnish (Ag_2S)	$2Ag + S \rightarrow Ag_2S$
Photosynthesis	Carbon dioxide gas (CO_2) + water (H_2O)	Glucose ($C_6H_{12}O_6$) + Oxygen gas (O_2)	$6CO_2 + 6H_2O \rightarrow C_6H_{12}O_6 + 6O_2$

Forming a compound always involves a chemical change, whereas forming a mixture involves only physical changes. Similarly, it requires a chemical change to break apart a compound. Mixtures, however, can be separated through physical changes. For example, saltwater can be boiled, thus leaving the salt behind; and blood can be spun in circles (centrifuged), thus separating it into its various components.

Chemical changes can be represented by chemical equations. A chemical equation uses formulas and numbers to keep track of a chemical change. The starting materials, called the *reactants,* are listed on the left side of the equation. The final materials, called the *products,* are listed on the right side. An arrow separates the two sides. The equation for the burning of coal, which is mostly carbon, would be written as:

$$C + O_2 \rightarrow CO_2$$

A chemist reads this equation as, "carbon plus oxygen yields carbon dioxide." In this reaction, carbon and oxygen are the reactants, and carbon dioxide is the product. Table 9-9 gives some examples of chemical changes.

Properties and Chemical Changes

The new substances produced by a chemical change have their own set of properties. These properties differ from those of the original substances that reacted, since those substances are no longer present as separate substances. For example, the element sodium is a soft metal that explodes on contact with water. The element chlorine is a poisonous, green gas. When sodium and chlorine combine in a chemical reaction, they produce sodium chloride, commonly known as table salt. The new substance formed has completely different properties from those of the original materials, which no longer exist as separate substances. During a chemical reaction, the atoms are rearranged to form new substances. This involves the breaking of existing chemical bonds and the formation of new bonds.

Both physical and chemical changes occur in nature. The wearing away of a mountain by streams is an example of a physical change called **erosion**. Erosion is the physical wearing away of rock material at Earth's surface. The Grand Canyon in Arizona was formed over millions of years by this physical change.

The Statue of Liberty in New York City is made of copper but does not look copper-colored. This is due to a chemical reaction between the copper and the air, which produces a new, green-colored substance (patina). The chemical wearing away of a metal is called **corrosion**. Corrosion, which forms a new substance, is a chemical change. Erosion, which only moves substances around, is a physical change.

Conservation of Matter

In a chemical change, no atoms are created and no atoms are destroyed. Every atom that is present before a reaction takes place is still there after the reaction takes place. What has changed is the way the atoms are arranged. Chemical reactions change only the way atoms are bonded to one another.

Figure 9-12 shows what happens when hydrogen and oxygen combine to form water in a chemical reaction. How many atoms of hydrogen are there before the reaction takes place? How does this compare with the number of hydrogen atoms after the reaction takes place? There are four hydrogen atoms before and after the reaction. How are the starting substances (the reactants) different from the substances formed (the products)?

Before the reaction, each hydrogen atom was bonded to one other hydrogen atom; after the reaction, each hydrogen atom was bonded instead to an oxygen atom. How would the mass of the starting materials compare with the mass of the materials formed? The mass remains the same, since no atoms were created or destroyed. This is an example of the **Law of Conservation of Matter**, which states that matter can neither be created nor destroyed in a chemical reaction. It can, however, be changed from one form to another. It is important to remember to account for all the substances before and after a reaction. In particular, it may be easy to forget about gases, which escape in the air, since it is difficult to capture and weigh them.

Figure 9-12. *The Law of Conservation of Matter: In a chemical reaction, such as the formation of water, there are the same numbers of atoms before and after the reaction.*

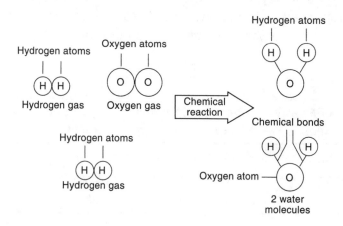

Energy and Chemical Changes

As you have learned, new substances are formed during a chemical change. An example is making table salt from sodium and chlorine. However, simply mixing sodium and chlorine together does not produce table salt. Energy is needed to start the chemical reaction. Likewise, a match does not start to burn until you strike it. The friction caused by striking the match provides the heat energy needed to start the chemical reaction of burning.

Many chemical changes must be started by the addition of energy, in the form of heat, light or electricity. However, some chemical changes do not require the addition of energy to get them started. The rusting of iron and the tarnishing of silver are examples of such reactions.

As a chemical reaction proceeds, energy is either absorbed or released. For example, the burning of a match releases energy in the form of heat and light. The chemical reaction that occurs in a battery releases electrical energy. On the other hand, when food is cooked, heat energy is absorbed by the chemical changes taking place. Table 9-10 gives some examples of chemical changes that absorb energy and chemical changes that release energy.

We can use chemical reactions to supply us with heat when we need it. For example, campers often use chemical hand warmers in cold weather. When they open the packet, the chemicals in the hand warmer react with oxygen in the air to release heat.

Reactions that absorb heat are also useful. A cold pack contains two chemicals that absorb heat when they react with each other. To start the reaction, you simply break the seal that separates the two chemicals.

Reversible Reactions

Many physical changes, such as melting and dissolving, can be reversed. Water can be turned into ice by chilling it, and ice can be turned back into water by warming it. Sugar can be dissolved in water, and the water can be evaporated to separate it from the sugar. These are examples of changes in which substances can be returned to their original form.

Most chemical changes, however, are very difficult to reverse. It is impossible to "unburn" a match. However, some chemical reactions are reversible. An important example is the reaction that takes place in a rechargeable battery. When you recharge a battery, you are reversing the chemical reaction used by the battery to produce electrical energy. Reversing a chemical change usually requires much more energy than does reversing a physical change.

Table 9-10. *Energy and Chemical Changes*

Chemical Changes That Release Energy	Type of Energy Released	Chemical Changes That Absorb Energy	Type of Energy Absorbed
Burning of wood	Light, heat	Cooking an egg	Heat
Battery starting a car	Electricity	Recharging a battery	Electricity
Decomposing of organic matter	Heat	Photosynthesis	Light

Making Predictions; Determining a Quantitative Relationship; Graphing Data

Nancy learned in chemistry class that temperature is a major factor in determining the rate of a chemical reaction. To investigate the effects of temperature on reaction rate, she decided to time a chemical reaction at several temperatures. Nancy's results are presented in the table below. Study the table and answer the following questions.

Temperature and Reaction Rate

Trial Number	Temperature	Time for Completion of Reaction
1	20°C	80 seconds
2	30°C	40 seconds
3	40°C	20 seconds

QUESTIONS

1. The data seem to indicate a trend: for each 10°C increase in temperature, the time needed to complete the reaction was
   (1) doubled
   (2) cut in half
   (3) decreased by 20 seconds
   (4) increased by 20 seconds

2. Assuming that the observed trend remains constant for all temperatures, how long would the reaction take at 50°C?
   (1) 15 seconds   (2) 5 seconds   (3) 10 seconds   (4) 20 seconds

3. Construct a graph by copying the numbered axes provided below onto a separate sheet of graph paper. Then plot the data from the table, as well as your answer to question 2, on the graph. To do this, mark a point at the intersection of a temperature line and a time line for the result of each trial. (The result of Trial 1 has been plotted as an example to guide you.) Finally, connect the points with a smooth curve.

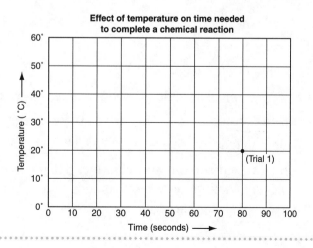

Effect of temperature on time needed to complete a chemical reaction

Rate of Reactions

As discussed earlier, milk turns sour when left out of the refrigerator. However, milk that is refrigerated eventually turns sour, too, though it takes longer to occur. The chemical reaction of souring, which happens rapidly at room temperature, occurs much more slowly in a cold refrigerator. In fact, most chemical reactions take place faster at higher temperatures. Frying an egg takes less time on a high flame than on a low flame. An increase in temperature increases the rate (speed) of a reaction.

Another factor that affects the rate of a chemical reaction is the size of the reacting particles. In general, the smaller the particles are, the faster the reaction is. For instance, a log burns more slowly than does an equal amount of sawdust.

The same factors that influence the rate of a chemical change also affect the rate of many physical changes. Recall that powdered sugar dissolves more rapidly than a cube of sugar does, and sugar dissolves more quickly in hot tea than it does in iced tea.

Review Questions

Multiple Choice

30. Which is only a physical change?

    (1) souring of milk   (3) melting of ice

    (2) burning of oil   (4) rusting of iron

31. Which process involves a chemical change?

    (1) photosynthesis   (3) freezing water

    (2) boiling water   (4) melting ice

32. A chemical change always

    (1) forms one or more new substances

    (2) absorbs heat

    (3) releases heat

    (4) absorbs electricity

33. In making an omelet, which process involves a chemical change?

    (1) melting butter   (3) frying eggs

    (2) chopping onions   (4) adding salt

34. A chemist mixed sodium and chlorine, but no reaction took place. A probable explanation for this outcome is that

    (1) the reaction only releases energy

    (2) the reaction only absorbs energy

(3) these substances cannot react

(4) energy must be added to start the reaction

35. During a chemical change,

(1) energy is always released

(2) energy is always absorbed

(3) energy is either absorbed or released

(4) energy is neither absorbed nor released

36. When making iced tea, David noticed that there was less ice after he mixed the ice with the hot tea. On observing this, David remembered that melting is a

(1) chemical change in which energy is absorbed

(2) chemical change in which energy is released

(3) physical change in which energy is absorbed

(4) physical change in which energy is released

37. Hydrogen gas is produced in a chemical reaction between zinc and an acid. Which setup would most likely have the fastest reaction rate?

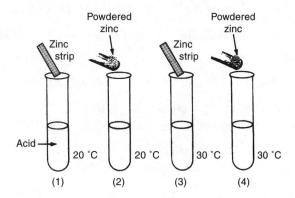

(1) zinc strip at 20°C   (3) zinc strip at 30°C

(2) powdered zinc at 20°C   (4) powdered zinc at 30°C

38. Food cooks faster at a higher temperature. This is because, as the temperature increases, the rate of a chemical reaction usually

(1) increases   (2) decreases   (3) remains the same

39. The diagram shows four samples of wood, each with a mass of one kilogram. Which sample would most likely burn the fastest?

(1) log   (2) planks of wood   (3) toothpicks   (4) sawdust

40. Which change would be the most difficult to reverse?

(1) melting an ice cube, because it is a physical change

(2) dissolving sugar in water, because it is a chemical change

(3) burning a match, because it is a chemical change

(4) rusting an iron nail, because it is a physical change

41. George cracks open an egg and finds that it has a very bad smell. The egg has become rotten. This change is best described as

(1) physical, because no new substances are formed

(2) physical, because a new substance was formed

(3) chemical, because no new substances are formed

(4) chemical, because a new substance was formed

42. A molecule of carbon dioxide (CO_2) is made of one atom of carbon and two atoms of oxygen. Carbon dioxide can best be classified as

(1) an element because it is made of one type of atom

(2) an element because it is made of two types of atoms

(3) a compound because it is made of one type of atom

(4) a compound because it is made of two types of atoms

43. Which subatomic particles are found outside the nucleus of an atom?

(1) positively charged protons   (3) negatively charged neutrons

(2) positively charged electrons   (4) negatively charged electrons

44. The diagram below represents an atom of helium. Based on this diagram, which statement is true?

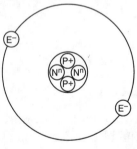

Helium atom

(1) The positively charged protons are found outside the nucleus.

(2) The positively charged protons are found inside the nucleus.

(3) The negatively charged protons are found outside the nucleus.

(4) The negatively charged protons are found inside the nucleus.

45. When 12 grams of carbon (C) react with exactly 32 grams of oxygen (O), carbon dioxide is formed. The mass of the carbon dioxide is

(1) greater than 44 grams   (3) less than 44 grams

(2) equal to 44 grams   (4) not determinable

Thinking and Analyzing

46. What is the difference between a physical and a chemical change? Give at least two examples of each.

47. Sulfur in insoluble in water while sugar is very soluble in water. How might you separate a mixture of sulfur, sugar and water?

To answer questions 48 and 49 refer to the charts below, which show two elements (zinc and sulfur) and their properties. The arrows indicate that the elements may be combined in a physical change to form a mixture or they can combine in a chemical change to form a compound.

Element	Properties		
Zinc	Good conductor of electricity, gray		
Sulfur	Nonconductor of electricity, yellow		

Physical change →

Properties of Zinc and Sulfur Mixture

Moderate conductor of electricity, yellow and gray

Chemical change →

Properties of Zinc Sulfide Compound

Nonconductor of electricity, white

48. What evidence indicates that a chemical change took place when the zinc and sulfur combined to form zinc sulfide?

49. During the chemical change a bright, white flash of light was observed. What does this indicate about the reaction between zinc and sulfur?

Chapter *10*

Motion and Machines

Points to Remember

▶ A force is a push or a pull. Balanced forces acting on an object cause the object to be at rest, and unbalanced forces acting on an object cause the object to move.

▶ Mass is a measure of the amount of matter in an object. The weight of an object is a measurement of the gravitational pull on the object. The amount of gravitational force produced depends on the mass of an object and the distance from its center to the center of the object that is pulling on it.

▶ Speed, velocity, and acceleration are ways of describing the motion of an object.

▶ Newton's first law states that an object at rest will remain at rest and an object in motion will remain in motion, unless a force acts on it. Newton's second law states that the acceleration of an object depends directly on the force applied and inversely on its mass. Newton's third law states that every action has an equal and opposite reaction.

▶ Work is done when an applied force moves an object. Machines make work easier The six simple machines are the lever, inclined plane, screw, pulley, wheel and axle, and wedge.

Force, Mass, and Motion

Force

(handwritten note: Mass - the amount of matter in an object)

A *force* is a push or pull (see Figure 10-1). To open a refrigerator door you pull the door. To move a computer mouse across a mouse pad you push or pull the mouse. To lift a log for the fireplace you must pull the log up against the force of gravity. Table 10-1 lists examples of pushing and pulling forces.

(a)   (b)

(handwritten note: I just drew that beautiful thing)

Figure 10-1. *Pulling (a) and pushing (b) forces illustrated by opening and closing a door.*

A force can also stop an object's motion, change its speed of motion, or change its direction of motion, as the following examples show.

1. Force stops motion: a falling acorn striking the ground; glove catching a baseball.
2. Force slows motion: friction slowing a skateboard; car going from a flat road to an uphill road.
3. Force changes direction: tennis racket striking a tennis ball; wind causing a fly ball to curve.

Table 10-1. *Pushing and Pulling Forces*

Pushing Forces	Pulling Forces
Hitting a baseball	Pulling a rope in a tug-of-war
Closing a refrigerator door	Opening a closet door
Moving a shopping cart	Lifting a shovel full of dirt
Hammering a nail	Climbing a rope
Wind knocking a tree down	Gravity pulling an apple to the ground

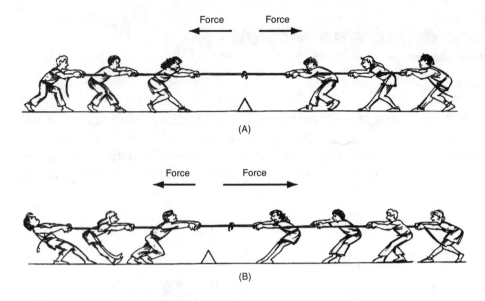

Force   Force

(A)

Force   Force

(B)

Figure 10-2. *A tug-of-war showing balanced forces (A) and unbalanced forces (B).*

When the forces acting on an object are balanced, the object is at rest. The teams in a tug-of-war (Figure 10-2A) do not move when the pulling forces are equal. However, balanced forces become unbalanced when an additional force is introduced. Figure 10-2B demonstrates that motion occurs when an additional force is introduced to one side of a balanced tug-of-war. Table 10-2 lists examples of balanced and unbalanced forces.

The Force of Gravity

All objects in the universe exert a gravitational pull on every other object. This is most noticeable on Earth's surface when you drop an object, and it falls to the ground. The amount of gravitational force between two objects is directly proportional to the masses of the objects and inversely proportional to the distance between the centers of the objects. That means the greater the masses the greater the gravitational force, and the further apart the objects the less the gravitational force. On Earth, an apple may fall from a tree and hit the ground. Earth is much more massive than the apple; thus, the apple appears to fall to Earth rather than Earth falling toward the apple. An apple falling from a branch 10 meters high takes about 1.4 seconds to

Table 10-2. *Balanced and Unbalanced Forces*

Balanced Forces	Unbalanced Force
Car parked in driveway	Car pulling out of driveway
Apple hanging on a tree	Apple falling to the ground
Roller coaster on top of a hill	Roller coaster coming downhill

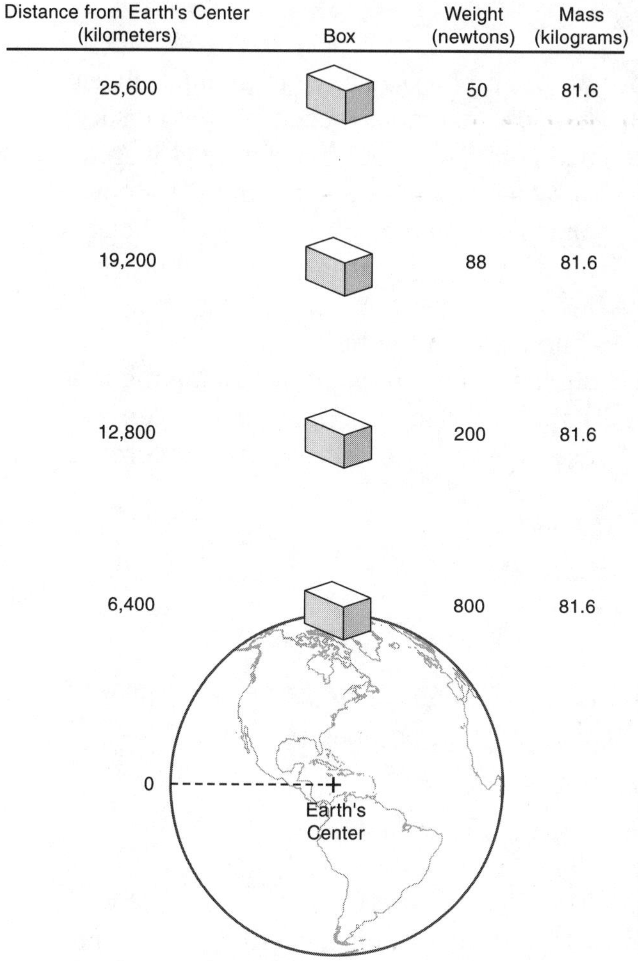

Distance from Earth's Center (kilometers)	Box	Weight (newtons)	Mass (kilograms)
25,600		50	81.6
19,200		88	81.6
12,800		200	81.6
6,400		800	81.6
0	Earth's Center		

Figure 10-3. *As an object moves away from Earth, its weight decreases, but its mass remains the same.*

reach the ground. The gravitational pull on the same apple 40,000 kilometers above Earth is much less. At this distance, it would take the apple much longer to fall 10 meters.

The weight of an object is a measurement of the gravitational pull exerted on the object. Gravitational force is expressed in newtons (metric units). Figure 10-3 shows an 800-newton box at different distances from Earth's center. The gravitational force decreases, and the weight of the box also decreases.

Mass

Mass is a measure of the amount of matter in an object. The mass of an object, unlike its weight, remains constant and does not change. Figure 10-3 shows that although the weight changes, the mass of the box is unchanged, as it is moves away from Earth's center.

A bowling ball and a baseball have different masses. A bowling ball contains more matter and is more massive than a baseball. The greater the mass of an object, the greater the force necessary to move the object. Thus, a greater force is needed to throw a bowling ball than to throw a baseball.

Motion

Motion is a change in the position of an object relative to another object, which is assumed to be at rest. Therefore, motion is recognized with respect to a nonmoving object or reference point. A school bus and its driver appear to be in motion when viewed by parents watching the bus drive down the street. The bus and driver change position relative to the nearby houses and trees. However, the bus driver does not appear to be in motion when viewed by the children on the bus. The driver does not appear to change position relative to the windows and seats on the bus.

Speed, *velocity*, and *acceleration* are three ways of describing the motion of an object. **Speed** is the distance traveled per unit of time. It can also be described as the rate of change in position of an object. The formula below is used to determine speed:

$$\text{speed} = \frac{\text{distance}}{\text{time}}, \text{ or } s = \frac{d}{t}$$

A car travels from Binghamton to Albany, a distance of about 200 kilometers in 2 hours. The car's average speed is determined by:

$$s = \frac{d}{t} = \frac{200 \text{ km}}{2 \text{ hr}} = 100 \text{ km/hr}$$

Some other units of labeling speed are meters/minute, centimeters/second.

Velocity and speed are similar. However, velocity also depends on direction; it has a directional component. **Velocity** is the speed of an object in a specific direction. The direction is assumed to be a straight-line direction such as North, East, Southwest, etc. The formula below is used to determine velocity.

$$\text{velocity} = \frac{\text{distance}}{\text{time}}, \text{ or } v = \frac{d}{t}$$

An airplane traveling from Detroit to New Orleans, a distance of 1600 kilometers in 2 hours, would determine its velocity by

$$v = \frac{d}{t} = \frac{1600 \text{ km}}{2 \text{ hr}} = 800 \text{ km/hr in a southwesterly direction.}$$

The units, km/hr, would be the same as that for speed.

Acceleration is the rate of change in velocity. (The change in velocity may be a change in direction.) Although acceleration refers to an increase or decrease in velocity, the term deceleration is commonly used to describe a decrease in velocity. Acceleration occurs when a car increases its velocity, and deceleration occurs when a car decreases its velocity. The formula below is used to determine the rate of acceleration.

$$\text{acceleration} = \frac{\text{final velocity} - \text{starting velocity}}{\text{time}}, \text{ or } a = \frac{v_f - v_s}{t}$$

A car increases its velocity from 10 kph to 30 kph in 2 seconds. Its acceleration is determined by:

$$a = \frac{v_f - v_s}{t} = \frac{30\,\text{kph} - 10\,\text{kph}}{2\,\text{sec}} = \frac{20\,\text{kph}}{2\,\text{sec}} = 10\,\text{kph/sec}$$

Other units of labeling acceleration are meters/minute/second, centimeters/second/second, etc.

Process Skill

How do you determine the average speed of a car trip?

Mike lives in Albany, New York. Last summer he took four car trips with his family. Mike's log of each trip indicated the distance and time it took to get to where they were going. The chart below summarizes the distance and time for each trip.

From/to	Distance (km)	Time (hr)	Average Speed
Albany to Buffalo	480	5	
Albany to New York City	240	2.4	
Albany to Cooperstown	120	1.5	
Albany to Lake Placid	200	2.8	

QUESTIONS

1. Using the formula s =d/t determine the average speed for each trip in km/hr units. Round your answer to the nearest tenth.
2. What factors affected the average speed of the trips, causing them to differ?

How do you determine acceleration of a moving car?

Ahh my best Friend hehe.

In a car trip between Binghamton and Rome, New York, Jessica recorded the car's velocity every minute for a period of 12 minutes. The chart shows her recorded times.

Time (minutes)	Velocity (kph)
1	70
2	70
3	70
4	80
5	90
6	100
7	100
8	90
9	85
10	80
11	75
12	70

QUESTIONS

1. Between the 1-minute mark and the 3-minute mark the car was
   (1) accelerating
   (2) decelerating
   (3) traveling at uniform velocity
   (4) stopped

2. Using the formula $a = \dfrac{v_f - v_s}{t}$ determine the acceleration of the car between the 3- and 6-minute marks.

3. Between what minute marks was the car decelerating?

Review Questions

Multiple Choice

1. Of the four different balls listed, the greatest force would be needed to move the
   (1) golf ball
   (2) baseball
   (3) basketball
   (4) bowling ball

2. The mass of the Washington Monument would be

(1) greatest in Washington, DC

(2) greatest at the North Pole

(3) greatest on the moon

(4) the same at all the above locations

Questions 3 and 4 refer to the following paragraph.

A force is a push or a pull. A force can start motion, stop motion, change the speed of motion, or change the direction of motion. In volleyball, the person serving the ball tosses it up to start the ball moving. The server then strikes the ball, changing its direction and increasing its speed toward the opposing team. When the ball reaches the other team, a player applies a new force causing the ball to slow and go up. Then, a player applies a stronger force to change direction and send the ball back to the serving team. This continues until the point is won and the motion of the ball is stopped.

3. Striking the ball when serving in volleyball is applying

(1) a pushing force

(2) a pulling force

(3) no force

(4) both a pushing and a pulling force

4. The purpose of striking the ball in a volleyball game is to

(1) place the ball in motion   (3) change the ball's speed

(2) change the ball's direction   (4) all of the above

5. Closing the front door of a house can be done with

(1) a pushing force

(2) a pulling force

(3) either a pushing or a pulling force

(4) neither a pushing nor a pulling force

6. A rocket traveling in space increases its velocity from 32,000 kph to 40,000 kph. To do this, the rocket must

(1) increase its mass   (3) increase the force acting on it

(2) decrease its mass   (4) decrease the force acting on it

Speed is determined by dividing the distance traveled by the time that it takes to travel the distance. The formula $s = d/t$ describes the relationship of the three factors. Using the formula answer questions 7 and 8.

7. An ant walked across one meter of sidewalk in one minute, what is the speed of the ant?

(1) 1 meter/second   (3) 1 meter/minute

(2) 10 meters/minute   (4) 1 kilometer/hour

8. If it took 1/2 the time to travel the same distance, how would the speed change?

    (1) double   (2) triple   (3) quadruple   (4) one-half

9. Speed and velocity are similar, except velocity has

    (1) an acceleration component   (3) a time component

    (2) a distance component   (4) a directional component

Thinking and Analyzing

Questions 10–12 refer to the following description.

Roberto threw a baseball up into the air. It reached a height of 9 meters and then started to come down. He then threw the ball to a height of 12 meters.

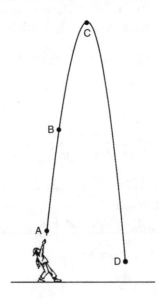

10. What force pulled the ball down?

11. What is the most likely reason that the ball went higher on Roberto's second throw? b/c he put more increased th

12. According to the diagram, at which point in the arc of the baseball's flight are the forces balanced?

Laws of Motion

Newton's Laws of Motion

In the mid 1660s, Sir Isaac Newton formulated the three laws of motion. These laws explain how forces affect the motion of all objects. Even today, Newton's laws of motion remain the basis for understanding the motion of objects.

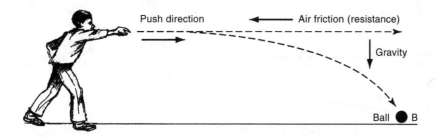

Push direction ← *Air friction (resistance)*

↓ *Gravity*

Ball ● B

Figure 10-4. *Air friction is the force that slows the ball and gravity pulls the ball back to Earth.*

First Law of Motion

The *first law of motion* states that **an object at rest will remain at rest and an object in motion will remain in motion, unless an outside force acts on the object.** There are two parts to this law. First, any object at rest will not move unless some force acts on it. An empty garbage can will remain at the curb or a dead leaf will remain on the front lawn until some force moves it. The force that moves the garbage can may be a person, and the wind may be the force that blows the leaf away. Second, any moving object will continue to move in the same direction, at the same speed, until a force acts on the object to change its speed or direction. A thrown baseball will move in a straight line at a constant speed until another force affects it. The force of air friction will decrease the speed of the ball, and the force of gravity will change the direction of the ball by pulling it down toward Earth (see Figure 10-4).

The tendency of an object at rest to remain at rest or an object in motion to remain in motion is called *inertia*. In other words, inertia is the tendency of an object to resist any change in its motion. The more massive an object is, the greater its inertia, or the greater it will resist a change in motion. When you are riding in a moving car that stops suddenly, your body continues to move forward. Also, when you are seated in a parked car that suddenly accelerates, you feel your body move backward. When a car stops, your moving body resists the stopping action. When a car accelerates, your body resists being put in motion (see Figure 10-5).

Figure 10-5. *The first law of motion: Objects resist a change in motion— an effect you can feel in a starting or stopping car.*

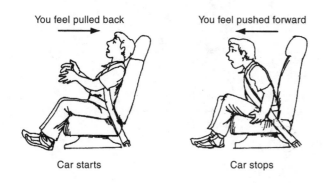

You feel pulled back →

You feel pushed forward ←

Car starts

Car stops

Second Law of Motion

The *second law of motion* states **the relationship among force, mass, and acceleration**. The law is commonly expressed by the formula:

$$\text{acceleration} = \frac{\text{Force}}{\text{mass}}, \text{ or } a = \frac{F}{m}$$

A large force acting on a given mass will cause a greater acceleration on that mass than a small force. For example, a mother (large force) pushing her husband (given mass) on a swing will cause a greater acceleration than their daughter (small force) pushing her father. (Figure 10-6A)

Also, a small mass acted on by a given force will have a greater acceleration than a large mass. A mother (given force) pushing her daughter (small mass) will cause a greater acceleration than the mother pushing the husband (large mass). (Figure 10-6B)

Figure 10-6. *(A) A large force pushing a given mass will cause a greater acceleration than a small force. (B) A small mass pushed by a given force will experience a greater acceleration than a large mass.*

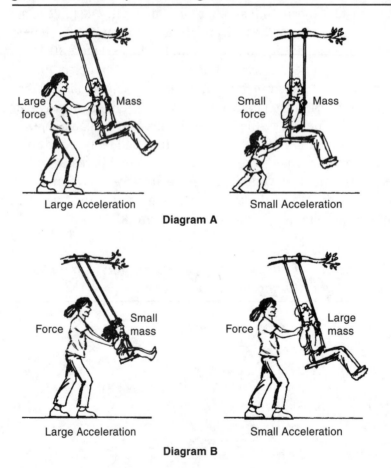

Diagram A

Diagram B

Laboratory Skill

Using Math to Analyze Data

Newton's second law of motion states the relationship among force, mass, and acceleration. This relationship is expressed by the following formula:

$$Force = mass \times acceleration, \text{ or } F = m \times a$$

The following chart shows six mathematical examples of how the formula is applied.

	Acceleration (meters/second/ second)	Force (newtons)	Mass (kilograms)
Example 1	1	1	1
Example 2	2	20	10
Example 3	1	10	10
Example 4	1	20	20
Example 5	0.5	10	20
Example 6	2	X	50

QUESTIONS

1. If the mass of a body remains the same and the force moving it is doubled, then the body's acceleration is doubled. Which two examples in the chart above demonstrate this?

2. If the force remains the same but the body is replaced with an object whose mass is twice that of the body, then the acceleration
   (1) is doubled
   (2) is halved
   (3) remains the same
   (4) is equal to mass

3. In Example 6 in the chart, the unknown force X would be
   (1) 50 newtons
   (2) 100 newtons
   (3) 150 newtons
   (4) 52 newtons

4. According to the chart, a newton is
   (1) a force that will move 1 kg at 1 meter/second/second
   (2) a force that will move 20 kg at 10 meters/second/second
   (3) a force that will move 10 kg at 20 meters/second/second
   (4) a measure of mass

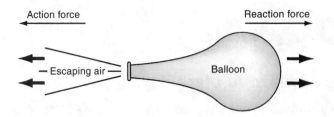

Figure 10-7. *The third law of motion: every action has an equal and opposite reaction.*

Third Law of Motion

The *third law of motion* states that **for every action there is an equal and opposite reaction**. A simple demonstration of blowing up a balloon and letting it go shows how this law works. When the air is released from the balloon, the balloon moves in the opposite direction (see Figure 10-7). Kicking a soccer ball, hot gases shooting out of a rocket engine, and walking are all actions that produce an equal and opposite reactive force.

 Review Questions

Multiple Choice

13. When you walk, your feet push against the ground, and the ground pushes back against your feet. This process demonstrates Newton's law, which states that

    (1) a body at rest remains at rest unless a force affects it

    (2) a body in motion remains in motion unless a force affects it

    (3) a large mass requires a large force to move it

    (4) every action has an equal and opposite reaction

14. When a golf ball is placed on a tee, it will remain there until the golfer strikes the ball. This demonstrates Newton's law, which states that

    (1) a body at rest remains at rest unless a force affects it

    (2) a body in motion remains in motion unless a force affects it

    (3) a large mass requires a large force to move it

    (4) every action has an equal and opposite reaction

15. Morris helped his dad push their car out of the garage to repair it. He pushed as hard as he could, but the car only moved very slowly. This example demonstrates the principle that

(1) a body at rest remains at rest unless a force affects it

(2) a body in motion remains in motion unless a force affects it

(3) a large mass requires a large force to move it

(4) every action has an equal but opposite reaction

16. On a breezy day, Esther pushed her model sailboat across the still water in a pond. According to Newton's first law, the sailboat should have continued across the pond and landed at Point A. However, it landed at Point B. The most likely reason for this was that

(1) the force of the wind changed the sailboat's direction of motion

(2) fish swimming near the sailboat changed its direction of motion

(3) the force of the moving water changed the sailboat's direction of motion

(4) the sailboat accelerated across the pond

17. A pitcher on the New York Mets exerts a pushing force on the baseball when delivering a pitch to a batter. The batter swings and misses the ball. The force that eventually changes the direction of the ball's motion is most probably

(1) air friction

(2) gravity

(3) the catcher's mitt

(4) the shortstop

18. Jane was standing inside a bus when the bus suddenly started moving. She immediately lost her balance and stumbled toward the back of the bus. This happened because

(1) Jane was standing in the back of the bus

(2) Jane was at rest and her body resisted a change in its motion

(3) Jane was walking through the bus

(4) the force of gravity was greater than the force of the moving bus

19. According to Newton's first law, when pushing a penny across a table, the penny will slide to the edge of the table and fall off. However, if the force applied is not strong enough, the penny will stop on the table. The force that stops the penny is

(1) gravity

(2) friction

(3) air

(4) the force of the table

Questions 20–23 refer to the following short paragraph.

A CO_2 cartridge was mounted on the top of a toy car with the nozzle of the cartridge pointing toward the rear of the car. When the compressed gas was released from the cartridge, the toy car moved 6 meters across the room and stopped. This was repeated using a car that had a mass ten times greater than that of the first car.

20. The reaction of the first car to the release of gas would be to move

(1) sideways (2) backward (3) forward (4) upward

21. The second car most likely moved

(1) more than 6 meters

(2) about 6 meters

(3) less than 6 meters

22. The reaction of the cars to the release of gas from the cartridge

(1) supports Newton's first law—a body in motion remains in motion and a body at rest remains at rest unless a force affects it

(2) supports Newton's third law—every action has an equal and opposite reaction

(3) supports both Newton's first and third laws

(4) does not support any of Newton's laws

23. The mass of the second car was greater than that of the first car, while the force affecting the second car was the same as the force affecting the first car. The acceleration of the second car

(1) was less than that of the first car

(2) was greater than that of the first car

(3) was the same as that of the first car

(4) cannot be compared with that of the first car with the available information

24. Newton's second law of motion states the relationship among force, mass, and acceleration is $F = m \times a$. According to the second law of motion, an increase in force on a given mass will cause

(1) a decrease in acceleration
(2) an increase in acceleration
(3) a decrease in the mass
(4) an increase in the mass

Thinking and Analyzing

A lawn sprinkler is shown in the illustration below. When the water is turned on, it comes out of the nozzle and sprays onto the lawn. Study the illustration and answer questions 25–27.

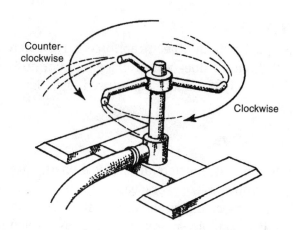

25. When the water comes out of the nozzle, in what direction will the sprinkler turn? *Clockwise*

26. What law governs the direction in which the sprinkler moves?

27. If the force of the water coming out of the nozzle is increased, what will happen to the sprinkler's speed of rotation? *It will increase in speed.*

Machines

Work

Scientists use the word "work" in a very specific way. They say that **work** is done when a force causes an object to move over a distance. The amount of work done depends on the amount of force applied and the distance the object is moved. The relationship among work, force, and distance is given by the formula:

Work = Force × distance, or $W = F \times d$

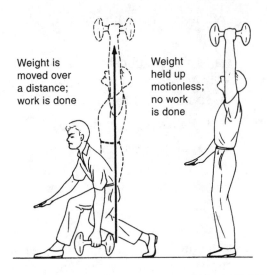

Weight is moved over a distance; work is done

Weight held up motionless; no work is done

Figure 10-8. *Work is done when a force acts over a distance.*

When a force is applied to an object, the force may or may not cause the object to move. If the force does not produce motion, no work is done. A force results in work only if motion is produced (see Figure 10-8).

Machines and Work

A *machine* is a device that transfers mechanical energy from one object to another object. Machines make work easier to perform. They do this by multiplying force, and by changing the direction or the distance over which a force is applied. For example, a wrench multiplies applied force when removing a tight bolt. A pulley can change the direction of a force when lifting a box. A loading ramp attached to the back of a truck reduces the force but increases the distance the object must be moved.

The force a machine has to overcome is called the *resistance*, and the force applied is called the *effort*. Using a machine can reduce the amount of effort needed to overcome a given amount of resistance. However, using a machine does not decrease the amount of work.

An example will make this clearer. Suppose you had to lift a box weighing 450 newtons up onto a platform 2 meters high. To lift the box straight up by yourself, you would need to apply 450 newtons of force over a distance of 2 meters. Using the formula $W = F \times d$, you get:

$$W = 450 \text{ newtons} \times 2 \text{ meters} = 900 \text{ newton-meters}$$

However, if you set up a rope and pulley system to change the direction and distance of the force required, you might have to pull in 9 meters of rope using only 100 newtons of force:

$$W = 100 \text{ newtons} \times 9 \text{ m} = 900 \text{ newton-meters}$$

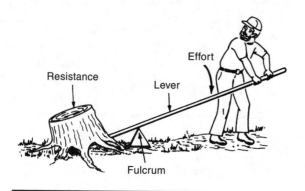

Figure 10-9. *A lever multiplies effort, making it easier to uproot a tree stump.*

The pulley system lets you use less effort over a longer distance to do the same amount of work.

Simple Machines

Machines help us do work faster and with less effort. There are six simple machines: the lever, pulley, wheel and axle, inclined plane, wedge, and screw. The pulley, and wheel and axle are modified forms of the lever, and the wedge and screw are modified forms of the inclined plane.

A *lever* consists of a rigid bar that can turn around a point called a *fulcrum*, as shown in Figure 10-9. Levers make work easier by multiplying the applied force. Examples of levers include a pair of pliers and a crowbar.

A *pulley* is a modified form of a lever. It can be used to change the direction of force or decrease the force needed to move a heavy object. At the repair shop, a pulley system is used to hoist a car's engine up and out the car. Figure 10-10 shows several types of pulleys.

Figure 10-10. *Three types of pulleys. (**R** stands for resistance, **E** stands for effort.)*

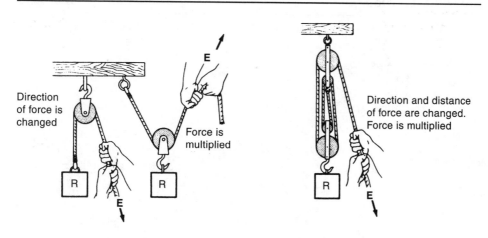

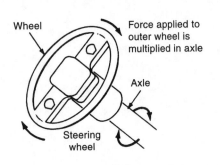

Figure 10-11. A steering wheel is an example of a wheel and axle.

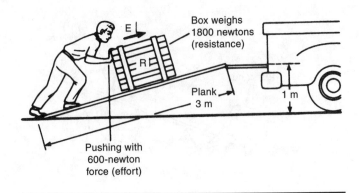

Figure 10-12. A loading ramp is an inclined plane.

A **wheel and axle** is also a modified lever; it consists of a large wheel with a smaller wheel, or axle, in its center. The wheel is attached to the axle so they turn together. When the outer wheel is turned, so is the axle (see Figure 10-11). Turning the outer wheel uses less force, but must turn a greater distance. Wheel and axle systems can be found in bicycles, car steering wheels, and doorknobs.

An **inclined plane** is a flat surface with one end higher than the other. A wheelchair ramp and truck ramp are inclined planes; so is a staircase. Figure 10-12 shows how an inclined plane makes work easier by decreasing the force needed and increasing the distance the force is applied.

The **wedge** is a double-sided inclined plane. The effort force is applied by driving the wedge into something, for example, an ax into a log. Other examples of wedges are knives, wood nails, and chisels. (Figure 10-13)

The **screw** is an inclined plane wrapped around a wedge or cylinder. Examples are wood screws, bolts, and car jacks. Basically, in using a wood

Figure 10-13. Three examples of wedges: (a) knife, (b) ax, and (c) wood nail.

Figure 10-14. The screw is based on the spiral inclined plane.

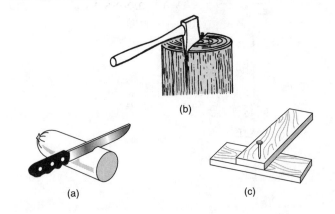

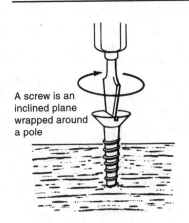

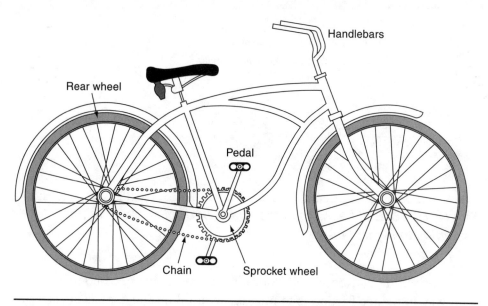

Figure 10-15. *Can you locate simple machines within a bicycle?*

screw, a circular force is applied to overcome the resistance of the wood and penetrate only a small distance. (Figure 10-14)

Many machines are called complex machines because they contain two or more simple machines. On an ordinary bicycle you should be able to locate a wheel and axle, a pulley system and a lever. (Figure 10-15 and Table 10-3)

Efficiency of Machines

Ideally, a machine's work output should equal the amount of work put into the machine. However, in reality, machines are never 100 percent efficient. The amount of work done by any machine is always less than the amount of work put into it. This is true because some of the work put into a machine is converted into heat energy, and thus wasted. The heat is produced by friction, which occurs where the machine's moving parts come in contact.

Returning to our earlier example, suppose you lift a 450-newton box up 2 meters, using pulleys. Although 900 newton-meters of work output was accomplished, you actually had to perform more than 900 newton-meters of work. Some of your work input is wasted because friction between the wheel and axle of each pulley and between the rope and each pulley creates heat.

Table 10-3. *Simple Machines on a Bicycle*

Wheel and Axle	Rear wheel, handlebars, sprocket wheel
Pulley	Chain
Lever	Pedals

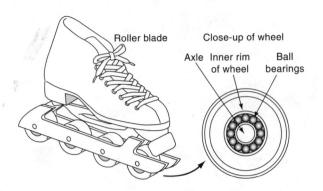

Figure 10-16. *Ball bearings in the wheel of a roller blade reduce friction as the wheel turns.*

A machine can be made more efficient by reducing friction. A common way to reduce friction is to put grease or oil on the contact surfaces of a machine's moving parts. Other methods include waxing the contact surfaces, sanding the surfaces to make them smoother, or using ball bearings between the surfaces (Figure 10-16).

Process Skill

Designing an Observation Procedure

Many common tools are levers of some kind. For example, scissors, shovels, and salad tongs are levers. Levers can be grouped into three basic classes, depending on where the lever encounters the *resistance*, where the *effort* is applied, and where the *fulcrum* (the point around which the lever turns) is located. Diagram 1 illustrates the three classes of levers.

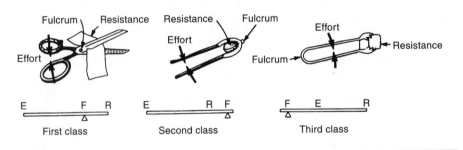

Diagram 1. *Three classes of levers.*

A first-class lever, such as a pair of scissors, has the effort (**E**) applied on one end, the resistance (**R**) on the other end, and the fulcrum (**f**) in between. A second-class lever, like a nutcracker, has the fulcrum and effort at opposite ends, and

(Continued)

the resistance in the middle. A third-class lever, such as a pair of ice tongs, has the resistance and the fulcrum at opposite ends, and the effort applied in the middle.

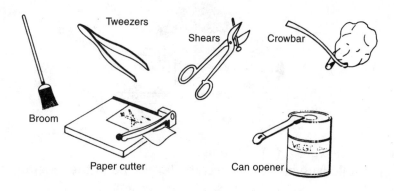

Diagram 2. *Examples of levers.*

Diagram 2 shows some examples of levers. Can you determine which lever class each item represents? Drawing a lever diagram for each one will help you do this. First, draw a line to represent the item. Next, think about how you use the item, and try to identify the positions of the fulcrum, effort, and resistance. Where does the item meet resistance? Is effort applied to one of the ends of the item, or somewhere in between? Where does the object turn or change direction? Once you have located the fulcrum, resistance, and effort, and labeled them on your lever diagram, you can then classify the item using the definitions of lever classes given above.

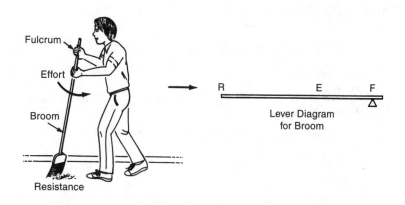

Diagram 3. *Using a broom, and its lever diagram.*

Here is an example of how this would be done for a broom. The lever diagram fits the definition of a third-class lever. Do the other items on your own, and then answer the following questions.

(Continued)

1. The wheelbarrow is a type of lever. The wheel in front is the location of the

   (1) effort   (2) fulcrum   (3) resistance
2. Which of the following items matches the lever type in the diagram below?

E   R   F

   (1) scissors   (2) paper cutter   (3) tweezers   (4) bottle opener

Review Questions

Multiple Choice

28. The force that a machine has to overcome is called
    (1) effort
    (2) resistance
    (3) friction
    (4) energy

29. The chain of a bicycle is greased in order to
    (1) increase weight
    (2) reduce air drag
    (3) reduce friction
    (4) increase resistance

30. A screw is a modified form of
    (1) wheel and axle
    (2) lever
    (3) pulley
    (4) inclined plane

31. Andrea moved a heavy box using a pulley system with an efficiency less than 100%. If her work output was 800 newton-meters, her work input was

(1) greater than 800 newton meters

(2) less than 800 newton meters

(3) exactly 800 newton meters

(4) no way to tell

32. The pulley on the flagpole in the illustration makes it easier to raise the flag by

(1) decreasing the amount of work required

(2) changing the direction of the force applied

(3) putting out more work than is put into it

(4) making the flag lighter

33. What type of simple machine shown below is being used to split the wood?

(1) wheel and axle

(2) pulley

(3) lever

(4) wedge

34. In science class, Mary tested the efficiency of four machines and recorded the results in a chart. For which machine must she have made an error?

Machine	Efficiency
Lever I	75%
Lever II	100%
Lever III	60%
Pulley	30%

(1) Lever I

(2) Lever II

(3) Lever III

(4) Pulley

Thinking and Analyzing

35. What three simple machines are being used in the diagram to help move bricks from position A to position B?

36. A student tries to lift a heavy box off the ground, but cannot make it move. Even though the student exerts great effort, no actual work is done. Explain.

Questions 37–40 refer to the following paragraph and the accompanying diagrams.

Lenno and Chin set up a ramp similar to the one in the diagram below to do an experiment. They attached a 40-newton metal block to a cord and a spring scale, as shown in the first diagram. During the experiment, they made a number of changes as they pulled the block up the ramp, as shown in the two other diagrams.

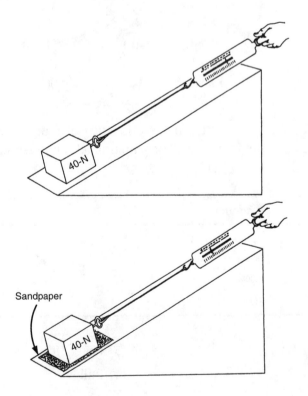

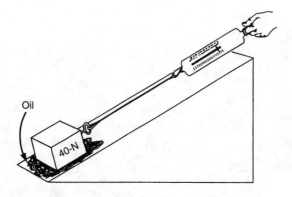

37. Which simple machine is illustrated by using the ramp to pull the weight?

38. When pulling the weight up the wooden ramp, the spring balance probably indicated an effort of

    (1) 40-N   (2) more than 40-N   (3) less than 40-N   (4) zero N

39. When Chin attached a sheet of sandpaper to the ramp and pulled the weight up across the sandpaper, how was the amount of effort required most likely affected?

40. When Lenno placed oil on the surface of the ramp and pulled the weight up across the oil, how was the amount of effort required affected?

Chapter *11*
Energy

Points to Remember

▶ Energy is the ability to do work. There are two states of energy: potential energy and kinetic energy. Energy occurs in the following forms: mechanical, chemical, nuclear, sound, heat, electric, and light.

▶ The Law of Conservation of Energy states that energy cannot be created or destroyed, it can only be transformed into other forms of energy.

▶ An electric circuit is a complete path for the flow of electricity. A circuit contains an electric source, a conducting path, and a device that uses the electricity.

▶ Sound and light travel in waves.

254

Energy

Energy is the ability to do work. *Work* is done when a force moves an object over a distance. A flowing river has the ability to move a boat. A car moving down the street can carry people from one place to another. Therefore, the river and the car possess some form of energy.

States of Energy

There are two basic states of energy: *potential* and *kinetic*.

1. *Potential energy* is stored energy that an object has because of its position above some reference level or its chemical composition. A rock on top of a cliff has potential energy because of its position above ground level. A lump of coal contains potential energy in its chemical bonds.

2. *Kinetic energy* is energy that an object has because it is moving. A rock falling off a cliff has kinetic energy. The heat given off by a burning lump of coal is also a form of kinetic energy. The faster an object moves, the more kinetic energy it has. Figure 11-1 shows examples of potential and kinetic energy.

Potential energy may be changed into kinetic energy when an object begins to move. Water held back by a dam has potential energy but no kinetic energy. Releasing the water and letting it flow changes its potential energy into kinetic energy.

Kinetic energy may also be changed into potential energy. When a ball is thrown straight up into the air, its kinetic energy of motion is changed

Figure 11-1. *Some examples of potential energy and kinetic energy.*

into potential energy as the ball rises higher above the ground. At the highest point of the ball's flight, it is motionless and has only potential energy. As the ball falls back to the ground, the potential energy changes back into kinetic energy.

Forms of Energy

Both potential and kinetic energy exist in many forms. For example, *mechanical energy* is the energy with which moving objects perform work. A hammer striking a nail, a jack lifting a car, and pedals turning the wheel of a bicycle are examples of things using mechanical energy. **Sound** is a type of mechanical energy. Table 11-1 shows some examples of different forms of energy.

Chemical energy is energy stored in certain substances because of the make up of their chemical bonds. When these substances are burned, the energy is released. Coal, oil, propane gas, and foods are examples of substances that contain chemical energy.

Nuclear energy is the energy stored within the nucleus (center) of an atom. This energy can be released by joining small nuclei together or by splitting large nuclei apart.

Heat energy is produced by the molecular motion of matter. All matter contains heat energy. Rubbing your hands together, cellular respiration, or burning fuel oil in a home heating system can produce heat energy.

Electric energy is produced by the flow of electrons through a conductor, such as a wire. Computers, light bulbs, and washing machines are all operated with electric energy. A generator produces electric energy.

Light is a form of radiant energy that moves in waves. Light as a form of energy can be demonstrated by using a magnifying glass to burn a hole in a leaf, or using a laser beam to burn a hole in a steel plate.

Conservation of Energy and Matter

The **Law of Conservation of Energy** states that energy can neither be created nor destroyed. However, energy can easily be transformed from one

Table 11-1. Different Forms of Energy

Form of Energy	Example
Mechanical	Fan
Sound	Bell
Chemical	Candle
Nuclear	Nuclear reactor
Heat	Toaster
Electric	Generator
Light	Lamp (bulb)

type of energy into one or more other types of energy. The *Law of Conservation of Matter* states that matter can neither be created nor destroyed. Energy and matter are related in such a way that they are interchangeable. That is, the total amount of energy and matter in the universe is constant, and each can be converted into the other. In the sun, large amounts of matter are being converted into light and heat energy. Scientists have been able to change matter into energy in nuclear reactors and energy into matter under special laboratory conditions.

Energy Transformations

Most of our daily activities involve the transformation of energy. For instance, when you take a bus to school each morning, chemical energy in gasoline is changed into mechanical energy that turns the wheels of the bus. At school, when the bell rings between classes, electric energy is transformed into sound energy. And at night, when you turn on a reading light, electric energy is changed into light energy. Figure 11-2 shows two common energy transformations.

Unusable Energy

Very often during the energy transformation process some heat energy is produced that is not usable. For example, a car's motor is designed to change chemical energy in the gasoline into mechanical energy to move the car. However, a running motor eventually becomes hot, due to the burning of fuel and the friction of the motor's moving parts rubbing against one another. In other words, some of the chemical energy is transformed into unusable heat energy instead of mechanical energy.

A vacuum cleaner is another example of a machine producing unusable energy. A vacuum cleaner contains a motor that transforms electric energy into mechanical energy. Run a vacuum cleaner a few minutes and you can feel that it gets warm. The electric energy entering the motor produces

Figure 11-2. *This hand-operated generator transforms mechanical energy into electrical energy, which is then transformed into light energy.*

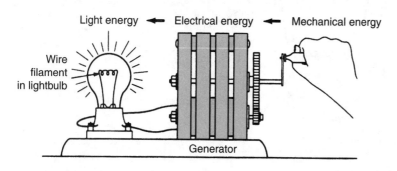

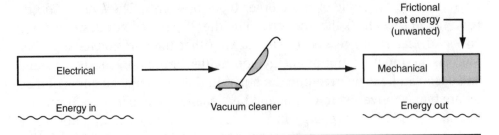

Figure 11-3. *Law of Conservation of Energy: "unusable" energy may be released during a transformation, but no energy is lost.*

mechanical energy to operate the appliance and an unusable amount of heat energy. This is illustrated in Figure 11-3.

Wave Characteristics

Sound and light are two forms of energy that travel in the form of waves. The waves produced by sound and light energy are similar to the waves produced by a pebble tossed into a calm pool of water.

The pebble entering the water is a source of energy and produces a series of waves that travel outward in all directions on the surface of the water. A wavy line, as shown in Figure 11-4 can represent these waves. The top of a wave is called the *crest*, and the bottom of the wave is called the *trough*. The distance from one point in a wave to a corresponding point in the next wave is called a **wavelength**. In other words, the distance from one wave crest to the next crest is a wavelength. The height of the crest or

Figure 11-4. *A representation of sound waves.*

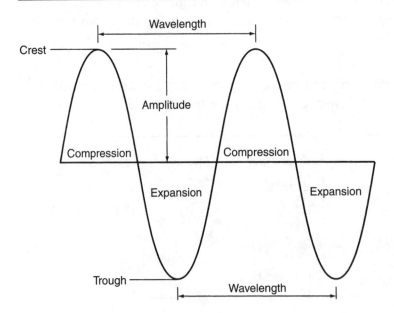

the depth of the trough of the wave measured from the undisturbed surface is the **amplitude** of the wave. The number of waves that pass by a fixed point in a given amount of time is called the wave's **frequency**. The substance through which the wave travels is called the *medium*. For instance, water is the medium for the waves produced by the pebble in the example above.

Types of Waves

A wave that vibrates up and down at a right angle to the direction in which the wave is traveling is called a *transverse wave*. Transverse waves can be demonstrated by using a rope tied to the back of a chair and shaking the untied end of the rope up and down. Although you see the wave moving along the rope, the actual material of the rope does not move forward, but rather moves up and down with each passing crest and trough.

A wave that vibrates back and forth within its direction of travel is called a *longitudinal wave*. A longitudinal wave can be demonstrated using a long coiled spring. Attach the spring to the back of a chair and stretch out the spring then push it in. You will see a series of "push-and-pull" waves pass through the spring. The area where the spring coils push close together is called a *compression*, and the area where the coils pull apart, or spread out, is called a *rarefaction*. The wavelength is measured from compression to compression. Sound waves are longitudinal waves. Figure 11-5 illustrates these two types of waves.

Figure 11-5. *A coiled spring can be used to demonstrate a longitudinal wave, and a rope can be used to demonstrate a transverse wave.*

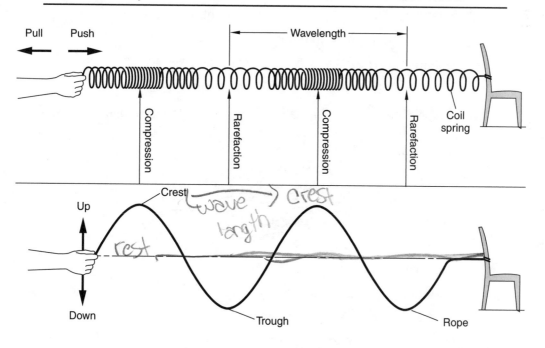

Review Questions

Multiple Choice

1. In the diagram below, a skier is about to start a slide (position A), ski down the hill (B), and stop at the bottom (C). At which position would the skier have the most kinetic energy?

    (1) A
    (2) B
    (3) C

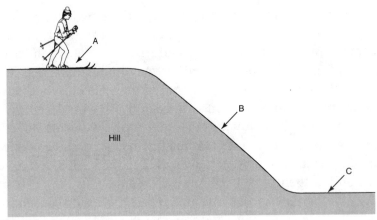

Hill

Questions 2 and 3 refer to the following activity.

Angie stretches a large rubber band and then releases one of its ends. The rubber band snaps back to its original shape.

2. When the rubber band was stretched, it had

    (1) potential energy   (3) no energy

    (2) kinetic energy   (4) motion energy

3. As the rubber band was snapping back, it had

    (1) potential energy   (3) no energy

    (2) kinetic energy   (4) position energy

4. What type of energy transformation is represented in the diagram below?

    (1) chemical energy to sound energy

    (2) sound energy to chemical energy

    (3) electric energy to sound energy

    (4) sound energy to electric energy

5. Sound energy is transformed to electric energy in the

   (1) telephone   (2) radio   (3) television   (4) hair dryer

6. Which item transforms 100% of the energy it receives into useful energy?

   (1) window fan   (3) battery-powered flashlight

   (2) hair dryer   (4) none of the items

7. While visiting an historic fort on vacation, John watched a demonstration of the firing of a cannon. When the cannon fired, the sound traveled to John and he heard the sound with his ears. The medium for the sound waves was

   (1) the cannon   (3) John's ears

   (2) the air   (4) none of the above

8. The amplitude of a sound wave indicates the loudness of a sound. The greater the amplitude, the louder the sound. The wave that represents the loudest sound is shown in

   (1)

   (2)

   (3)

   (4)

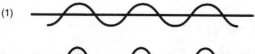

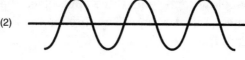

Questions 9–10 refer to the diagram below, which represents a transverse wave.

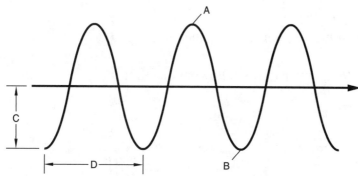

9. The crest of the wave is

   (1) Point A   (3) Distance C

   (2) Point B   (4) Distance D

10. The wavelength of the wave is

   (1) Point A   (3) Distance C

   (2) Point B   (4) Distance D

Thinking and Analyzing

Questions 11 and 12 refer to the following activity. In an effort to break a strip of aluminum metal, Charles bent it back and forth many times. He was surprised to notice the aluminum getting warm at the point of the bend.

11. This is an example of a transformation of

    (1) heat energy to mechanical energy

    (2) chemical energy to mechanical energy

    (3) chemical energy to heat energy

    (4) mechanical energy to heat energy

12. The heating of the metal strip in this case

    (1) is a gain of energy

    (2) is the creation of energy

    (3) is the loss of energy

    (4) is an unusable energy transformation

13. The diagram below represents three appliances that transform electric energy into mechanical energy. The three items that best fit into boxes A, B, and C are the

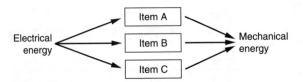

    (1) vacuum cleaner, table saw, fan

    (2) television, fan, refrigerator

    (3) radio, clock, fan

    (4) lawn mower, hair dryer, generator

Questions 14–16 refer to the following diagram.

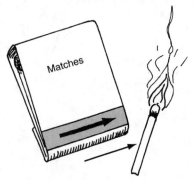

Lighting matches in a matchbook illustrates several different energy transformations.

14. When you strike a match on the rough strip of a matchbook cover, you transform

(1) mechanical energy to heat energy

(2) mechanical energy to chemical energy

(3) mechanical energy to electric energy

(4) heat energy to chemical energy

15. The burning match involves the transformation of

(1) chemical energy to light and sound energy

(2) chemical energy to light and heat energy

(3) chemical energy to heat energy, only

(4) heat energy to light energy, only

16. The matches in the matchbook contain

(1) potential chemical energy

(2) potential light energy

(3) potential heat energy

(4) kinetic heat energy

17. What type of energy is produced by each of the items listed below?

a. toaster

b. window fan

c. vacuum cleaner

d. leaf blower

e. telephone

f. can opener

g. hair dryer

h. lawn mower

i. chain saw

j. stove

Heat

Heat Energy

Heat is a form of energy produced by the vibrating motion of molecules. All matter is composed of molecules that are constantly vibrating. When heat is added to a substance, the molecules move faster and farther apart. When heat is removed from a substance, the molecules move slower and come closer together.

The addition of heat causes most substances to expand, and the removal of heat causes most substances to contract. Adding heat to a substance causes the molecules to move farther apart which causes it to increase its size. Bridges, railroad tracks, and sidewalks have expansion spaces. This allows them to expand and contract freely in response to the great temperature

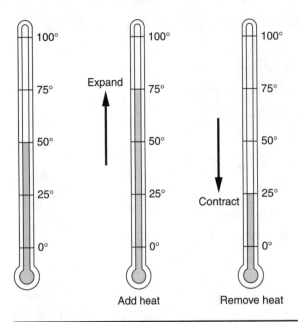

Figure 11-6. *When the alcohol in a thermometer is heated, it expands and indicates a higher temperature. When heat is removed, the alcohol contracts and indicates a lower temperature.*

changes that occur between summer and winter. Some thermometers contain alcohol that expands and contracts inside a glass tube, providing a way to measure temperature (see Figure 11-6).

Although most substances expand when heated and contract when cooled, water is an exception. In liquid form, when water cools, it contracts. However, when water is cooled to 4°C, it starts to expand and continues expanding until it becomes ice at 0°C. This is why an unopened bottle of water or soda will crack if left outdoors during freezing weather. The force generated by the expansion of water changing to ice is so powerful that it can crack glass, rocks, concrete, and even steel.

Heat Transfer

When a difference in temperature exists, heat travels from warmer objects or places to cooler objects or places. The tendency is for heat to become equally distributed. Heat can move by conduction, convection, or radiation.

Conduction is the transfer of heat by direct molecular contact. Metal objects conduct heat well. A metal spoon placed in a cup of hot tea quickly gets hot because heat is easily transferred from the tea to the spoon and from molecule to molecule within the spoon. A metal pot on a hot stove quickly distributes the heat throughout the pot by conduction. On the other hand, materials that do not transfer heat well can be used to insulate or reduce the flow of heat. A pot holder reduces heat flow so you can grasp the metal handle of a hot frying pan without being burned (see Figure 11-7).

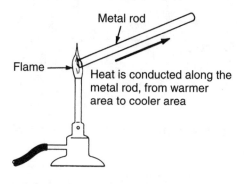

Metal rod

Flame —

Heat is conducted along the
metal rod, from warmer
area to cooler area

Figure 11-7 *Heat transfer by conduction.*

Convection is the transfer of heat by the flowing action within a liquid or gas. Warm air added to a room rises, and the cooler air sinks. This causes a circular flow called a convection current that distributes the heat within a room. Figure 11-8 demonstrates how a convection current occurs in a room and eventually equalizes the temperature.

Radiation is the transfer of heat through space in the form of waves. The heat from the sun reaches Earth in the form of waves that travel through the vacuum of space. Place your hand several inches below a light-bulb and you can feel the warmth of the bulb. The heat from the bulb reaches your hand by radiation.

Figure 11-8. *Home heating systems use convection currents to heat a room.*

![check mark] **Review Questions**

Multiple Choice

18. Water differs from most other substances in that it

    (1) expands when heated

    (2) contracts when cooled

    (3) expands when it freezes

    (4) contracts when it freezes

19. Placing a metal spoon with a temperature of 20°C into a cup of water with a temperature of 90°C will cause the spoon to

    (1) increase in temperature

    (2) decrease in temperature

    (3) remain the same temperature

    (4) contract in size

Metal spoon

20. A thermos bottle keeps warm liquids warm and cold liquids cold. A thermos bottle must be made from a good

    (1) heat conducting material

    (2) heat insulating material

    (3) heat expanding material

    (4) heat contracting material

21. A thermometer works on the principle that, when heated, a substance will

    (1) contract   (3) remain the same size

    (2) expand   (4) give off light

22. Heat from the sun reaches Earth by

    (1) conduction   (3) radiation

    (2) convection   (4) none of the above

Thinking and Analyzing

23. Which graph best shows the relationship between the temperature of a substance and the motion of the particles in the substance when it is heated?

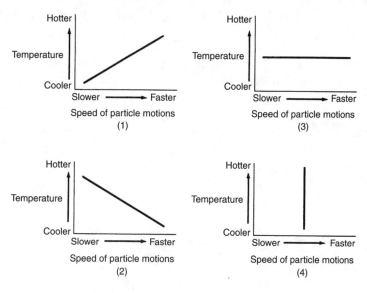

Hotter

Temperature

Cooler

Slower ⟶ Faster
Speed of particle motions
(1)

Hotter

Temperature

Cooler

Slower ⟶ Faster
Speed of particle motions
(3)

Hotter

Temperature

Cooler

Slower ⟶ Faster
Speed of particle motions
(2)

Hotter

Temperature

Cooler

Slower ⟶ Faster
Speed of particle motions
(4)

24. The end of a metal bar is placed in a flame for five minutes. The temperature is measured by thermometers at four points on the bar, as shown in the diagram below. The lowest temperature will most likely be recorded at thermometer:

(1) A

(2) B

(3) C

(4) D

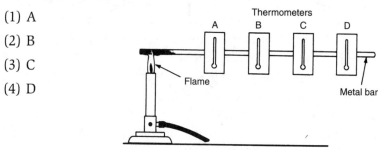

Thermometers

A   B   C   D

Flame

Metal bar

25. Explain how you can make the colored water in the glass tube rise and explain how you can make it fall.

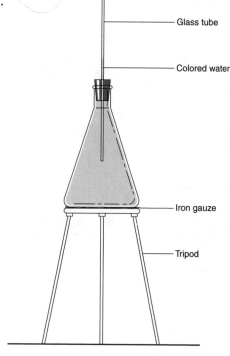

Glass tube

Colored water

Iron gauze

Tripod

26. Describe the movement of heat in a glass of soda when several ice cubes are placed in it.

27. A cloud forms by the process of condensation. The gaseous water changes into liquid water droplets. The temperature of the surrounding air

    (1) increases

    (2) decreases

    (3) remains the same

Magnetism and Electricity

Magnetism

A magnet has the ability to attract items made of iron, cobalt, or nickel (see Table 11-2). Bar magnets have their greatest strength located at the ends of the bar in areas called poles. Magnets have a north (N) pole and a south (S) pole of equal strength. If you cut a magnet in half, each half will have a north and a south pole. Hang a bar magnet from a string tied around its middle, and it will align itself with Earth's magnetic field. The pole of the magnet toward the north direction is the north pole of the magnet, and the pole toward the south direction is the south pole of the magnet (see Figure 11-9). A compass is a device that uses a magnet to help us determine directions.

When the ends of two magnets are brought together a push or pull force occurs. The law of magnetic poles states that like poles repel and unlike poles attract. If two N poles or two S poles are brought together, you feel a pushing force as they repel each other; however if a north and a south pole are brought together you feel a pulling force as they attract each other. The closer the magnets the greater the force (see Figure 11-10).

Around each magnet there is a magnetic field (see Figure 11-11 on page 270). Place a sheet of paper over a magnet and sprinkle iron fillings on the paper, and you can easily see the shape of the field. The iron fillings are most concentrated at the poles showing that the lines of force are greatest at the poles, but can also be seen around the whole magnet.

Table 11-2. List of Some Magnetic and Nonmagnetic Items

Magnetic	Nonmagnetic
Paper clip	Eraser
Staples	Pencils
Iron filings	Pieces of paper
File cabinet	Chalk
Carpenter's nail	Plastic pen

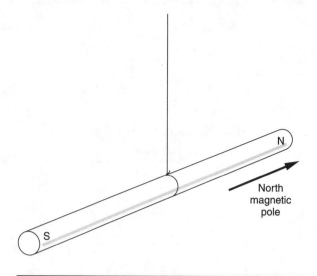

Figure 11-9. *A freely hanging bar magnet will align with Earth's magnetic field.*

Static Electricity

Atoms normally contain an equal number of electrons (negative [−] particles) and protons (positive [+] particles). Therefore, they are neutral. If additional electrons are added, the atom becomes negatively charged, and if electrons are removed, the atom becomes positively charged. Objects contain many atoms, so they can gain or lose many electrons. The whole object can become negatively charged if it has additional electrons, and positively charged if it has lost electrons.

When a glass rod is rubbed with a silk cloth, electrons are transferred from the glass rod to the silk. When a rubber rod is rubbed with a wool cloth,

Figure 11-10. *Like poles of a magnet repel, and unlike poles attract.*

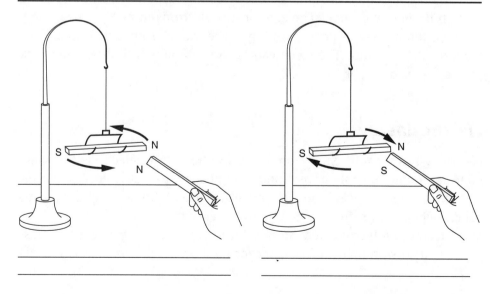

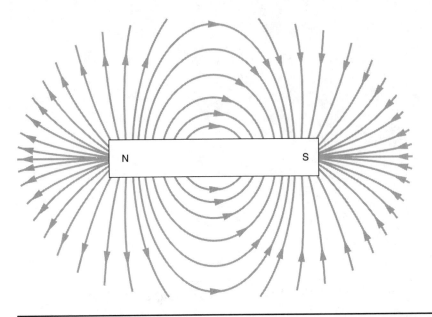

Figure 11-11. *The magnetic field around a magnet is concentrated around the poles.*

electrons are transferred from the wool to the rod. The surface of these substances receive an electric charge called static electricity. This is the same type of static electricity that causes a shock when you walk across a wool or nylon rug and touch a metal object. On a much larger scale, lightning is also a form of static electricity.

The law of electric attraction and repulsion states that items with like charges repel each other and items with unlike charges attract each other. Hang two balloons from the ceiling so they just touch each other. Rub each with a wool or flannel cloth. Each balloon now has a negative charge, and they repel each other. Rub a glass rod with a silk cloth. Bring the positively charged glass rod toward a negatively charged balloon, and the balloon is attracted to the glass rod (see Figure 11-12).

Another example of static electricity can be demonstrated by running a comb through your hair. The comb gains electrons to become negatively charged. Rip a sheet of paper into tiny pieces about 1 cm across and place them on the tabletop. The electrically charged comb will attract and pick up the pieces of paper.

Electric Energy

Electricity is a form of energy produced by the flow of electrons from one point to another. People have found many uses for electric energy. For instance, the electricity in your home can be used to power lightbulbs, air conditioners, television sets, and many other appliances.

Electric ***conductors*** are materials through which electricity moves easily. Conductors provide an easy path for the flow of electrons. Most metals

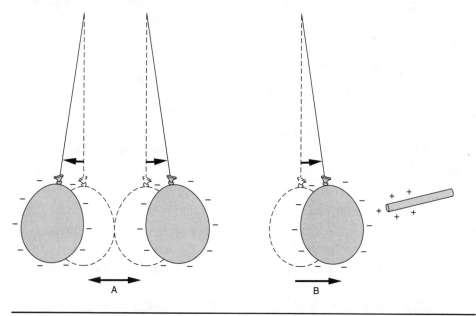

Figure 11-12. (A) Negatively charged balloons will repel, and (B) a negatively charged balloon is attracted to a positively charged glass rod.

are good conductors. Substances that are poor conductors and do not allow electrons to flow through them are called **insulator**s. Rubber, glass, and plastic are common insulators.

Electric wires in your home consist of a conductor, like copper or aluminum, wrapped in a protective coating of an insulator, like rubber or plastic (see Figure 11-13). These wires provide a safe path for electricity.

Production of Electricity

An electric current is produced when electrons move through a conductor, such as a copper wire. The two most common methods of producing electricity are *electromagnetic induction* and chemical action.

When a magnet is pushed through a coiled copper wire, it causes or *induces* a flow of electrons through the wire. This process is known as **electromagnetic induction** (see Figure 11-14 on page 272). Generators (see Figure

Figure 11-13. An electric wire consists of a conductor (metal) wrapped in an insulator (plastic).

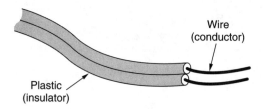

Wire
(conductor)

Plastic
(insulator)

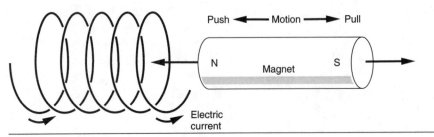

Figure 11-14. *An electric current is produced when a magnet is pushed and pulled through a coiled wire.*

11-2 on page 257) are devices that spin a coiled wire within a magnetic field to produce electricity. In a generator, the mechanical energy of a spinning *turbine* is transformed into electric energy for human use. The turbine may be powered by running water or wind, or by burning coal or oil. The electricity in our homes, schools, and office buildings is produced primarily by this method.

A battery is a device that converts chemical energy into electric energy. Certain chemicals, such as zinc and carbon, produce a flow of electrons when placed in a conducting material (see Figure 11-15). If the conducting material is sulfuric acid, as in a car battery, the battery is called a *wet cell battery*. If the conducting material is a solid, as in a flashlight battery, it is called a *dry cell*. Batteries are used in cars, portable radios and clocks, laptop computers, and many other devices.

An ***electric circuit*** is a complete path for the flow of electricity. A circuit must contain a source of electric energy, a conducting path for the flow of electrons, and a device that uses the electric current. The source of electricity could be a battery or a generator. Wires usually provide the path for the electricity in a circuit, and the device could be a lamp or a CD player.

At least two wires are needed for a complete circuit, one wire to carry a flow of electrons to the device and another to carry electrons back to the

Figure 11-15. *Cross-section of a dry cell battery.*

Figure 11-16. *A simple electric circuit, with an on-off switch included.*

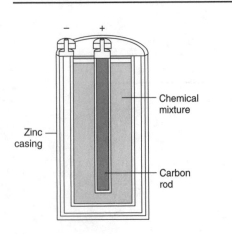

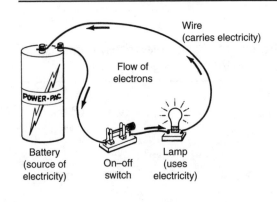

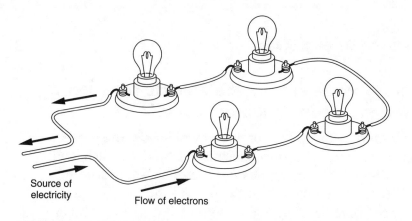

Source of electricity

Flow of electrons

Figure 11-17. *A series circuit showing electricity flowing through each lightbulb in a single, continuous path.*

source of electricity. In addition, a circuit often includes a switch that turns the flow of electricity on and off. Figure 11-16 shows a simple electric circuit with an on–off switch.

There are two types of electric circuits—*series* and *parallel.*

A **series circuit** has a single path of electricity through devices attached to it (see Figure 11-17). The same amount of electric current flows through each of the devices. Some colored holiday lights are wired in a series circuit. If any of the lightbulbs are removed or burn out, the electric path is broken, and all the bulbs go out.

A **parallel circuit** has two or more devices wired so that the electricity flows through a separate branching path to each device (see Figure 11-18). The current divides into each branch, so that the amount of current decreases as more branches are added. In a string of lightbulbs connected in a parallel circuit, if any of the bulbs are removed or burn out, the electricity continues to flow through the remaining bulbs, so they stay lit.

Figure 11-18. *A parallel circuit showing electricity branching to flow through each lightbulb.*

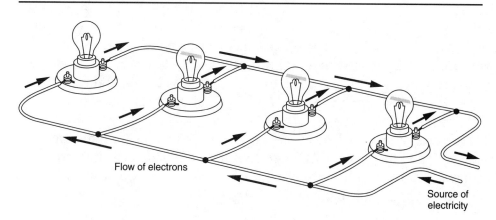

Flow of electrons

Source of electricity

How can you determine the electric conductivity of a material using a simple circuit?

A simple circuit can be used to measure the conductivity of solids. The diagram shows an electric circuit containing a source of electricity, a bulb, a switch, and wires with alligator clips. Strips or bars of various test materials are easy to hold with the clips. To test an item, you would attach the clips to the item, close the switch, and note whether the bulb lights or not. If the bulb lights, the item conducts electricity, and if the bulb does not light the item is a nonconductor.

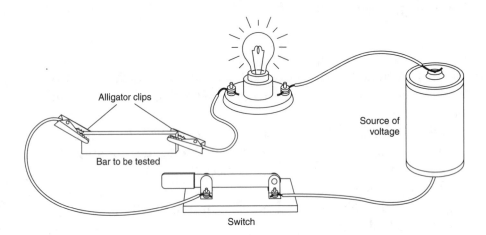

Metals such as silver, copper, and aluminum are good conductors. They allow electrons to pass through them easily. Nonconductors, such as rubber and plastic, do not allow the passage of electrons. Nonconductors are used as insulators on the outside of electric wires and plugs.

QUESTIONS

1. A student used a simple circuit to test the conductivity of the items listed below.

   (1) Plastic comb   (5) Brass screw
   (2) Glass stirrer   (6) Wooden splint
   (3) Plastic stirrer   (7) Copper wire
   (4) Aluminum foil   (8) Pencil

   Which of these items would most likely conduct electricity?

2. Two unknown types of wire (wire A and wire B) were tested with the conductivity apparatus. When wire A was used, the lightbulb was bright, and when wire B was used, the lightbulb was dim. Explain what this tells you about the conductivity of the wires.

3. The metal wires on home appliances are covered to protect us from the electric current carried in the wires. This cover does not conduct electricity, and is called an insulator. Which substance is commonly used as a wire insulator?

   (1) plastic   (2) copper   (3) aluminum   (4) glass

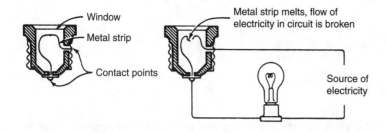

Figure 11-19. *How a fuse works. When the electric circuit begins to overheat, the metal strip in the fuse melts, breaking the flow of electricity.*

Safe Use of Electricity

Electricity is dangerous and can injure or kill living things. Though it has become a common part of our lives, electricity should always be handled carefully.

When too many appliances are using electricity from one circuit at the same time, the circuit can become overloaded. This overload makes the wires heat up, which can cause an electric fire. *Fuses* and *circuit breakers* prevent overloading of circuits by automatically interrupting the flow of electricity when it reaches a dangerous level. These protective devices are essential in a home wiring system. Figure 11-19 shows how a fuse works.

Grounding of electric appliances is another important safety precaution. Sometimes an electric charge can build up in an electric device. A person touching the device may receive a severe shock from this excess charge. Grounding helps prevent such accidents.

To ground an electric device, a wire is run from the device to a conducting material, such as a metal pipe, that is in contact with the earth. This provides a safe outlet for any excess charge in the device by conducting it to the earth, where it is absorbed. Some appliances come equipped with a three-pronged plug (see Figure 11-20). The third prong provides grounding and should always be used.

You should observe common-sense safety rules whenever using electricity. Always unplug electric devices before cleaning or repairing them. Electric wires should not be used if they are broken or frayed, and plugs should be replaced if the wire is broken near the plug. Never handle electric appliances while you are in water or if you are wet, because water can act as a conductor.

Figure 11-20. *A three-pronged plug coated with a plastic insulator.*

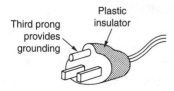

Multiple Choice

28. When the north poles of two magnets are brought close together, they

 (1) attract each other

 (2) have no effect on each other

 (3) repel each other

 (4) first repel then attract each other

29. The ends of a magnet are called

 (1) electrodes   (2) poles   (3) the field   (4) domains

30. Michael walked across the carpet and grabbed the doorknob. He got a small electric shock when he touched the metal doorknob. What caused the electric shock?

 (1) dampness in the air

 (2) faulty electric wiring

 (3) a transfer of heat

 (4) movement of electrons

31. A good conductor of electricity is

 (1) plastic   (2) rubber   (3) aluminum   (4) glass

32. The simple electric circuit shown below is incomplete. What item is missing?

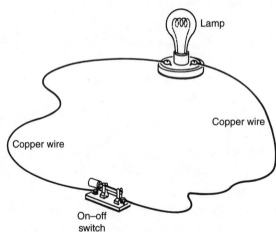

 (1) a conducting path for the electricity

 (2) a source of electric energy

 (3) a device to use the electricity

 (4) a switch to start and stop the flow of electricity

33. Electricity is

 (1) the flow of protons

 (2) the flow of electrons

 (3) the flow of atoms

 (4) the flow of molecules

34. A generator can transform mechanical energy into electric energy. It spins a coiled wire within a magnetic field to produce the electricity. This is called

(1) magnetism

(3) static electricity

(2) electromagnetic induction

(4) chemical electricity

Thinking and Analyzing

35. Explain why a house's electric circuit is a parallel circuit rather than a series circuit.

To answer questions 36 through 38, read the following sentences and study the diagram. James produced an electric circuit display for the school science fair. He connected 3 bulbs and 2 switches in the arrangement shown in the diagram.

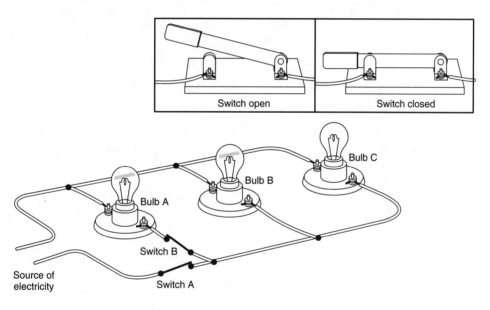

36. The setup put together by James is

(1) a parallel circuit

(2) a series circuit

(3) a combined parallel and series circuit

(4) not a circuit

37. If switches A and B are closed, then

(1) lightbulbs A, B, and C are on

(2) lightbulbs A, B, and C are off

(3) lightbulbs B & C are on, and bulb A is off

(4) lightbulbs B & C are off, and bulb A is on

38. If switch A is closed and switch B is open, then

  (1) lightbulbs A, B, and C are on

  (2) lightbulbs A, B, and C are off

  (3) lightbulbs B & C are on, and bulb A is off

  (4) lightbulbs B & C are off, and bulb A is on

39. Explain what would happen if a positively charged glass rod was brought close to a negatively charged balloon hanging from a string.

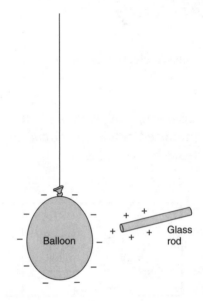

40. Two bar magnets are placed on the tabletop as shown in the diagram. Magnet B is slid slowly toward magnet A. Describe what will happen to magnet A?

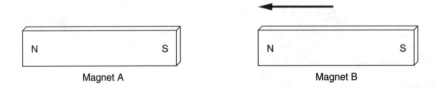

Magnet A   Magnet B

Sound

Sound Energy

Sound is a form of energy produced by a vibrating object. When an object vibrates, it moves rapidly back and forth. This motion pushes and pulls the surrounding air, producing alternating compressed and expanded layers of air particles called **sound waves** (see Figure 11-21). These sound waves spread outward in all directions from their source, somewhat like the circular ripples that are produced when you toss a pebble into a calm pool of water.

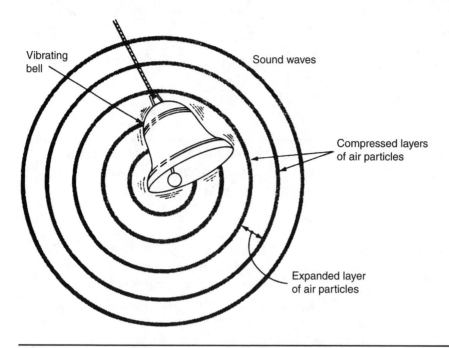

Vibrating bell

Sound waves

Compressed layers of air particles

Expanded layer of air particles

Figure 11-21. *A vibrating object produces sound waves.*

Objects that can produce sound include bells, radio speakers, guitar strings, or anything else that can vibrate. For instance, the sound of your voice is caused by the vibrating vocal cords in your throat. If you place your hand on your throat while you speak, you can feel the vibrations that produce the sound.

To hear a sound, it must be transmitted from a source to your ear. The substance that sound travels through is called the medium. The medium can be a solid, liquid, or gas. Most sounds we hear travel through air, a gas. Sound waves cannot travel through a *vacuum* because there are no particles of matter in a vacuum to transmit the sound.

Sound Waves

Although sound waves are longitudinal waves, they can also be represented as transverse waves (see Figure 11-5 on page 259). The crest of each wave represents the compressed-particle portion of the wave (the compression), and the trough represents the expanded-particle portion (the rarefaction). The larger the amplitude of the wave, the louder the sound.

Wave frequency is commonly expressed as the number of vibrations per second. Humans can hear sounds that range from 20 to 20,000 vibrations per second. A normal speaking voice ranges between 100 to 1000 vibrations per second and has a wavelength of 0.3 to 3.5 meters. The pitch describes how high or low the sound is. A high-frequency sound has a high pitch, and a low-frequency sound has a low pitch. A violin produces a high-pitched sound, and a bass drum produces a low-pitched sound.

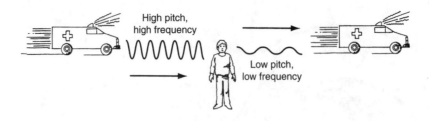

High pitch,
high frequency

Low pitch,
low frequency

Figure 11-22. *The Doppler Effect: When the source of sound moves toward the listener, the frequency heard increases; and when the source moves away form the listener, the frequency heard decreases.*

To a motionless observer, the frequency of sound waves produced by the horn on a moving train or the siren on an ambulance increases or decreases, depending on whether the sound source is moving toward or away from the listener. This change in sound frequency, called the **Doppler effect**, is recognized as a change in pitch. When a blaring ambulance siren is moving toward you, the sound waves are crowded together, producing a higher frequency and a higher pitch. When the siren is moving away from you, the sound waves spread out, producing a lower frequency and a lower pitch (see Figure 11-22).

The Speed of Sound

The speed of sound depends primarily on the density of the substance, or medium, through which it is passing. The denser the medium, the faster the sound waves can travel through it. Generally, sound travels fastest through solids, which have the greatest density, and slowest through gases, which have the least density. Table 11-3 gives the speed of sound through several substances. To a lesser extent temperature also affects the speed of sound.

Although the speed of sound can vary, it is always much slower than the speed of light. During a thunderstorm, for instance, a lightning bolt produces a flash of light and a clap of thunder at the same time. The speed of light is so fast that the light reaches us almost instantly. The sound of the thunder travels much more slowly, so we usually hear the thunder a few seconds after we see the lightning.

Table 11-3. *Speed of Sound Through Different Substances (at 25° C)*

Medium	State	Speed (m/sec)
Iron	Solid	5200
Glass	Solid	4540
Water	Liquid	1497
Air	Gas	346

How does temperature affect the speed of sound in air?

Sound travels at different speeds through different substances. The speed of sound is faster in solids, such as stone or metal, and slower in liquids and gases, such as water and air. The speed of sound through air is also affected by air temperature. The graph below shows the relationship between air temperature and the speed of sound. Study the graph and answer the following questions.

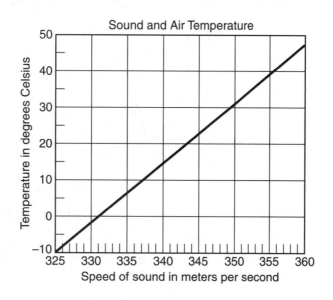

Sound and Air Temperature

QUESTIONS

1. Sound travels at a speed of 340 meters per second at about which temperature?
   (1) 22°C   (2) 10°C   (3) 6°C   (4) 15°C
2. At a temperature of 31°C, sound travels at about
   (1) 345 meters per second
   (2) 350 meters per second
   (3) 355 meters per second
   (4) 340 meters per second
3. What does the graph suggest about the relationship between air temperature and the speed of sound?
   (1) As air temperature decreases, the speed of sound increases.
   (2) As air temperature increases, the speed of sound remains the same.
   (3) As air temperature increases, the speed of sound increases.
   (4) As air temperature increases, the speed of sound decreases.

Multiple Choice

41. Sound is produced by

(1) expansions

(2) contractions

(3) reflections

(4) vibrations

42. Sound travels in waves that can be represented as shown in the diagram below. The top of a wave is called the crest, and the distance from crest to crest is called the wavelength. The diagram shows

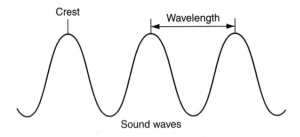

(1) one wave

(2) five waves

(3) three waves

(4) an incomplete wave

43. When using the apparatus in the diagram below, the student could not hear the ringing bell after the air was pumped out of the bell jar. This demonstrates that sound

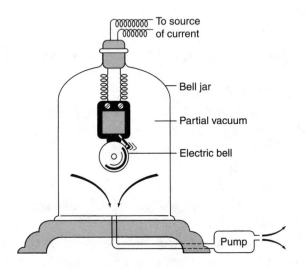

(1) can travel through the glass bell jar

(2) cannot travel through the glass bell jar

(3) cannot travel through a vacuum

(4) can travel through a vacuum

44. A passing train blows a whistle at a constant pitch as it approaches, crosses, and leaves a road crossing. A person standing by the road crossing hears the pitch of the whistle rise as the train approaches, and then get lower as the train passes by. This is caused by the

(1) wavelength

(2) Doppler effect

(3) amplitude of the sound

(4) different whistles on the train

Thinking and Analyzing

Read the following short paragraph and study the diagram. Then answer questions 45–47.

Sound travels about 340 meters/second in air. Josh yelled "hello" across a canyon. The sound traveled across the canyon and returned as an echo. Josh heard the echo two seconds after he yelled "hello."

45. The distance the sound traveled was about

(1) 340 meters   (3) 1360 meters

(2) 680 meters   (4) 170 meters

46. The distance across the canyon was about

(1) 340 meters   (3) 1360 meters

(2) 680 meters   (4) 170 meters

47. If the canyon had been 1020 meters across, Josh would have heard the echo in about

(1) 5 seconds   (3) 3 seconds

(2) 6 seconds   (4) 12 seconds

48. The bar graph below shows the average speed of sound through solids, liquids, gases, and a vacuum. Based on the graph, through which medium would sound travel fastest?

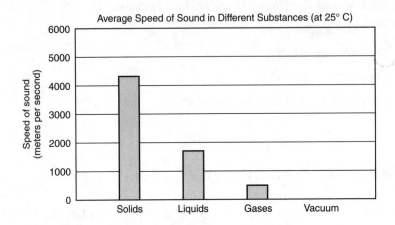

Average Speed of Sound in Different Substances (at 25° C)

(1) air   (2) rock   (3) water   (4) vacuum

49. What can cause sound waves to increase their speed?

50. Describe why someone can hear you calling them if they are in another room that is not directly in line with you.

Light

Light Energy

Light is a visible form of energy. Like sound, light travels in waves that move outward in all directions from its source. Light waves travel in straight paths called rays (see Figure 11-23). This is why objects block light rays and cast shadows. Light can travel through a vacuum, something sound cannot do. Light from the sun travels through the vacuum of space to reach Earth. The speed of light is extremely fast, about 300,000 kilometers per second. That is almost a million times faster than the speed of sound! Light that

Figure 11-23. *Light waves travel in straight paths. Shadows are produced when an object blocks their path.*

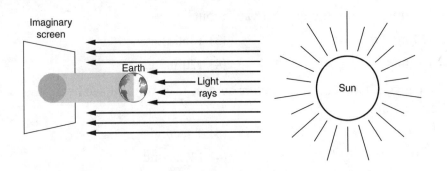

travels over distances we commonly encounter on Earth arrives in just a very small fraction of a second.

The sun is our main source of light energy. Fire and lightning are other sources of natural light. Light can also be produced artificially, as it is in a lightbulb.

Light Can Be Reflected, Absorbed, or Transmitted

When light strikes the surface of an object, three things can happen to the light rays as shown in Figure 11-24. Light rays may bounce back, or reflect, off the surface. Light rays may be absorbed and be transformed to heat energy. And light rays may be transmitted, passing through the object. A shiny, metal surface reflects most of the light that strikes it. Much of the light that strikes a blacktop road is absorbed and transformed to heat. Clear glass allows most light that strikes it to be transmitted through it.

We see objects because they emit light or their surfaces reflect light, and the light rays enter our eyes. The smoother the surface, the more accurate the reflection. A mirror gives an accurate reflection because it has a smooth, shiny surface. A wall does not produce a reflection, because its rough surface scatters the light.

Colored objects absorb light to varying degrees. Dark-colored objects absorb more light, which becomes heat energy, than do light-colored objects, which reflect more light. For this reason, people usually wear light-colored clothing to keep cool during hot, sunny weather.

Materials also differ in their ability to transmit light. Transparent materials, such as window glass, permit almost all of the incoming light to pass directly through it. Translucent materials, such as wax paper, let some light pass through, but scatter the light rays so the images are not transmitted clearly. Opaque materials, like wood and iron, do not allow any light to pass through.

Lenses

Sometimes, when light passes from one transparent substance into another, such as from air into water, the light rays are bent, or *refracted*. This is why a pencil in a glass of water looks broken or bent where it enters the water

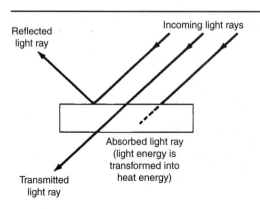

Figure 11-24. *When light strikes a surface light may be reflected, absorbed, and/or transmitted.*

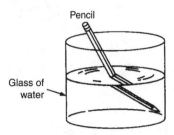

Pencil

Glass of
water

Figure 11-25. *Bending of light rays as they pass from one medium to another makes the pencil look bent or broken.*

(see Figure 11-25). The light rays being reflected from the pencil are refracted as they pass from the water into the air. This fact has been put to use in the making of lenses.

A *lens* is a piece of transparent glass or plastic that has curved surfaces. The curved surfaces refract light rays that pass through the lens. The shape of a lens determines how it bends light, as shown in Figure 11-26. A lens with surfaces that curve outward (convex) bends light rays so that they are focused in toward a common point. A lens with surfaces that curve inward (concave) bends light rays so that they spread out.

Images of objects seen through lenses may be larger, the same size, or smaller than the object itself. For instance, the lens of a camera forms smaller images of objects. A simple photocopy machine has a lens that forms images the same size as the original object. Binoculars contain lenses that magnify objects, making them appear larger.

The Electromagnetic Spectrum

Light waves are part of a larger group of energy waves that can travel through a vacuum at the speed of light. These are called *electromagnetic waves*, which include radio waves, microwaves, infrared waves, visible light, ultraviolet rays, X rays, and gamma rays. Together, these energy waves form a continuous band of waves called the electromagnetic spectrum (see Figure 11-27).

Figure 11-26. *The shape of a lens determines how it bends light.*

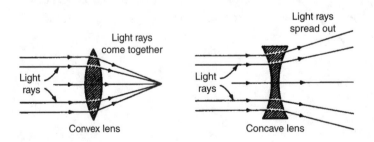

Light rays
come together

Light
rays

Convex lens

Light rays
spread out

Light
rays

Concave lens

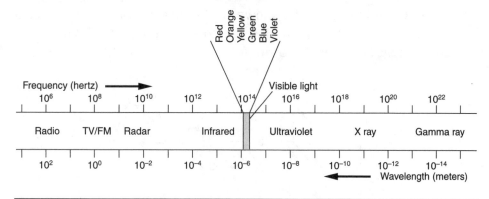

Figure 11-27. *The electromagnetic spectrum.*

Electromagnetic waves affect our daily lives in many ways. Radio waves are used for radio and television broadcasting. Microwaves are used in communications and microwave ovens. X rays are used to diagnose illnesses and injuries.

Ordinary white light makes up only a narrow range of the electromagnetic spectrum. White light is further divided into the color spectrum, which is based on wavelength. The colors of the spectrum, arranged in order of wavelength are red, orange, yellow, green, blue, and violet. Red has the longest wavelength and violet the shortest wavelength. The color that you see when you look at an object is the result of the wavelengths of light it reflects and absorbs. For example, a blue car looks blue because it reflects the blue wavelengths of light and absorbs all other wavelengths.

Overexposure to some electromagnetic waves can be harmful to living things. We should be especially careful to limit our exposure to electromagnetic radiation, such as X rays and ultraviolet rays, which have been linked to genetic mutations and cancer.

 Review Questions

Multiple Choice

51. A blast of dynamite set off by a roadwork crew produced a bright flash of light and a loud explosion. A person standing two kilometers away, with a clear view of the work site, would

    (1) hear the sound first, then see the flash

    (2) see the flash first, then hear the sound

    (3) see the flash and hear the sound at the same time

    (4) hear the sound, but not see the flash

52. When using the apparatus shown in the diagram below, the student could see the flame only if all three holes were lined up. What property of light does this demonstrate?

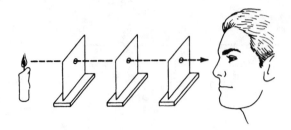

(1) Light rays are reflected from smooth, shiny surfaces.

(2) Light rays are absorbed as heat by dark-colored surfaces.

(3) Light rays travel in straight paths.

(4) Transparent objects transmit most of the light that strikes them.

53. We can see most objects because they

(1) bend light   (3) absorb light

(2) reflect light   (4) transmit light

54. The accompanying diagram shows three ways that light can behave when striking a sheet of colored glass. The light rays at location C have been

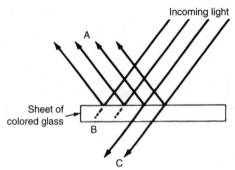

(1) reflected by the glass   (3) absorbed by the glass

(2) transmitted through the glass   (4) blocked out by the glass

55. When white light strikes a green wall,

(1) all of the light is absorbed

(2) all of the light is absorbed except green light, which is reflected

(3) all of the light is reflected

(4) all of the light is reflected except green light, which is absorbed

56. Sound waves and light waves are similar in that

(1) both can bend around objects

(2) both originate from an energy source

(3) both travel at the same speed

(4) both are a series of compressions and rarefactions

57. Lenses can produce images that are

    (1) larger than the original object

    (2) the same size as the original object

    (3) smaller than the original object

    (4) all of the above

58. X rays are often used for

    (1) radio and television broadcasting

    (2) communications and cooking food

    (3) diagnosing illnesses and injuries

    (4) all of the above

59. Overexposure to ultraviolet light from the sun

    (1) has no effect on humans

    (2) always has a good effect on humans

    (3) can have a harmful effect on humans

    (4) cannot occur

Thinking and Analyzing

Questions 60 through 62 refer to the following information.

Robert made a series of observations during a thunderstorm. He recorded the time difference between each flash of lightning and the thunder that followed it. The chart below shows his findings.

Lightning Bolts	Time Difference to Thunder (in seconds)
A	12
B	9
C	6
D	3
E	5
F	8
G	10
H	15

Robert knows that light traveling over relatively short distances arrives almost instantly, and that sound travels at about 340 meters/second in air. Robert estimated that for every three-second difference in time between a flash of lightning and the thunder, the lightning bolt was about one kilometer away. From the information in the chart, Robert could draw some conclusions.

60. The lightning bolt that was closest to Robert was

(1) B   (3) D

(2) C   (4) H

61. The lightning bolt that was about 2 kilometers (2000 meters) from Robert was

(1) C   (3) G

(2) E   (4) H

62. From the trend of the data, Robert concluded that the thunderstorm

(1) moved away from him and then toward him

(2) moved toward him and then away from him

(3) was continuously moving away from him

(4) was continuously moving toward him

63. Each of the three diagrams represents a light ray striking a surface. (a) Which diagram represents a mirror? (b) Which diagram represents a glass window?

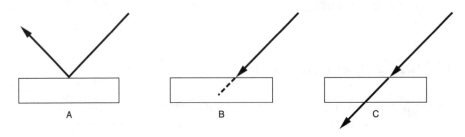

Use the diagram below to help you answer questions 64–66.

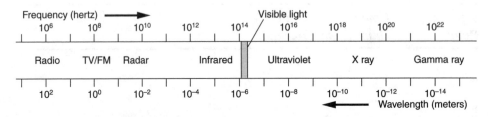

64. Using the electromagnetic spectrum describe two differences between X rays and radio waves.

65. What is the relationship between wave frequency and wavelength?

66. Which type of electromagnetic wave has a frequency of 10^{10} hertz and a wavelength of 10^{-2} meters?

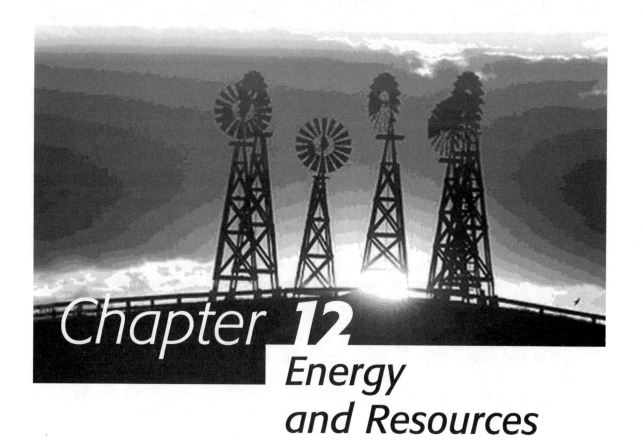

Chapter 12
Energy and Resources

Points to Remember

▷ Fossil fuels, our main sources of energy, come from the remains of dead plants and animals. Fossil fuels are nonrenewable resources.

▷ Burning fossil fuels produces gases such as carbon dioxide and other pollutants, which may affect Earth's atmosphere.

▷ Other sources of energy include hydroelectric, nuclear, solar, and wind. Each energy source has advantages and disadvantages.

▷ It is important to conserve our energy supply. Methods of conserving energy include reducing consumption, recycling materials, and reusing materials.

Forms of Energy

Everything that occurs in the universe involves energy. As you have learned, energy is the ability to do work—to make something move. Heat, light, sound, and electricity are all forms of energy. Humans have learned to describe, explain, and measure energy and to harness it for their use.

Measuring Energy

To compare the amounts of energy stored in various substances, we need some unit of measurement. The energy in foods and fuels can be measured and compared using a unit called the **calorie.** One calorie is the amount of heat energy needed to raise the temperature of one gram of water by one degree Celsius.

When describing the energy in food, we use the word Calorie spelled with a capital "C." This "food Calorie" is equal to 1000 ordinary calories, or 1 kilocalorie (kilo means one thousand). Calories indicate how much energy you can get from various foods. Digested food containing energy that is not needed by the body is stored, usually as fat. When you go on a diet, you count Calories to make sure you don't eat more food than your body needs for energy.

The rate at which energy is used (the amount of energy used over a certain period of time) can also be measured. The rate at which electric energy is used is measured in a unit called the **watt.** We can use watts to compare the rates at which different electric devices use energy. For instance, a 100-watt lightbulb uses twice as much electricity each second as a 50-watt bulb. Table 12-1 lists some common electric devices and their wattage.

Energy Consumption

Energy consumption is constantly increasing. The world's growing populations and economies require more and more energy. Fuels are sources of energy. As the demand for energy increases, so does the demand for fuel. We

Table 12-1. Some Electric Devices and Their Wattage

Electric Device	Wattage
Hair dryer	1200 watts
Lightbulb	100 watts
Electric shaver	7 watts
Small air conditioner	860 watts
Microwave oven	750 watts
Stereo	240 watts
Toaster oven	1400 watts

need more fuel to cook meals, heat homes, run industries, and power cars, ships, trains, and airplanes. We also use more fuel to produce electricity.

Electricity has become essential to our society. It is used for many purposes, such as heating and cooling buildings, running machines and appliances, and providing lighting. Electric energy can be transmitted easily over conductors such as metal wires. However, electric energy must be produced from other energy sources.

Fossil Fuels

The main energy sources used to produce electricity are **fossil fuels.** They are called fossil fuels because they were formed from the remains of plants and animals that lived and died long ago. Over time, these organic remains were changed into energy-rich substances. The most commonly used fossil fuels are oil (also called *petroleum)*, coal, and natural gas.

Oil is a sticky black liquid usually found trapped within rock layers deep underground. **Coal** is a black rock that occurs in layers, or seams, between other rock layers. **Natural gas** is commonly found underground with oil deposits. Each of these fossil fuels can be burned to provide energy for the production of electricity. Figure 12-1 shows the relative amounts of oil, coal, natural gas, and other energy sources used to produce electricity in the United States.

Other Uses of Fossil Fuels

Fossil fuels have many other uses besides producing electricity. Gasoline, used to power cars, and heating oil, used to heat homes, are both fossil fuels made from petroleum. Natural gas is used to heat homes and industries, and for cooking. (Gas stoves use natural gas.) Coal is used in industrial processes, such as the making of steel.

Fossil fuels are used to make many other important substances. Oil, in particular, has many such uses. Plastics, fertilizers, certain drugs, and synthetic fabrics like nylon and polyester are all products made from petroleum.

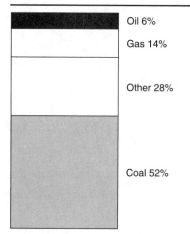

Oil 6%

Gas 14%

Other 28%

Coal 52%

Figure 12-1. Relative percentages of energy sources used to produce electricity in the United States.

Interpreting a Graph

A pictograph represents numbers, or quantities, by using pictures. In the accompanying diagram, each picture of a barrel represents one million barrels of oil. For example, the category "Industry and electricity" is represented by three barrels, which means that three million barrels of oil are used each day for these purposes.

Daily Uses of Oil

Light fuels and chemicals

Gasoline

Jets, trains, and diesel fuels

Heating oil

Industry and electricity

Fertilizers, tar, and grease

= One million barrels

Lost

How much oil is used daily for "Jets, trains, and diesel fuel"? Looking at the row of barrels for that category, you see there are two complete barrels plus part of a third barrel. This part is about one-quarter of a complete barrel, so it represents one-quarter of a million barrels of oil. This makes the total daily use of oil for "Jets, trains, and diesel fuel" equal to 2 1/4 million barrels. Use the diagram to help you answer the following questions.

QUESTIONS

1. Which products use the least amount of oil each day?
   (1) light fuels and chemicals   (3) heating oil
   (2) gasoline   (4) fertilizers, tar, and grease
2. About how much oil is lost every day?
   (1) half a million barrels
   (2) less than one-quarter of a million barrels
   (3) one million barrels
   (4) more than one-quarter of a million barrels
3. How might the graph differ in summer and winter?

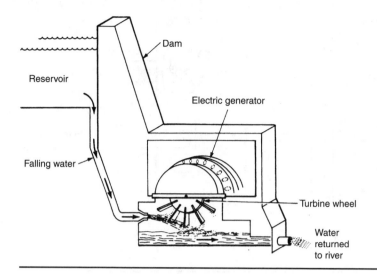

Figure 12-2. *A hydroelectric dam used to produce energy.*

Other Sources of Energy

While most of the energy we consume comes from fossil fuels, several other energy sources are currently in use. The most important of these are *hydroelectric energy* and *nuclear energy.*

1. **Hydroelectric energy** is electricity produced by the power of flowing water. Water in motion has kinetic energy, which can be transformed into electric energy. When water flows swiftly downhill at a waterfall or a dam on a river, it can be used to operate a *generator,* which produces electricity. A generator contains a *turbine,* a device similar to a paddle wheel on an old-fashioned steamboat (see Figure 12-2). The moving water turns the turbine, which spins a coil of wire inside an electromagnet in the generator. This creates an electric current.

   Hydroelectric energy is usually clean and inexpensive. Niagara Falls is a major source of electric energy in New York State. Figure 12-3 compares the energy sources used by electric companies in

Figure 12-3. *A comparison of energy sources used to produce electricity in Ohio and in New York.*

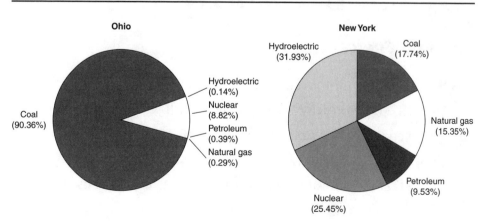

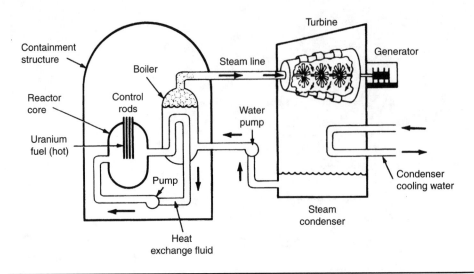

Figure 12-4. *A nuclear power plant: heat from a nuclear reaction changes water into steam, which turns turbines to produce electricity.*

New York with those used in Ohio. Notice that New York uses a much greater percentage of hydroelectric energy than does Ohio, while the percentage of coal used in New York is much smaller. The choice of energy sources varies from one state to the next, as different states have different sources available to them.

2. *Nuclear energy* is the energy stored in the nucleus of an atom. When this energy is released, it creates heat, which can be used to produce electricity. Nuclear power plants use **uranium** (a radioactive element found in certain rocks) as their fuel source. Uranium atoms are naturally unstable and can be split apart readily to release heat energy. The heat created by the uranium fuel in a nuclear reactor is used to boil water, thereby producing steam. The steam turns turbines that generate electricity. This is illustrated in Figure 12-4.

 ## Review Questions

Multiple Choice

1. Which is *not* a form of energy?

    (1) heat   (2) electricity   (3) gasoline   (4) light

2. The energy obtained from foods is generally measured in

    (1) Calories   (2) watts   (3) volts   (4) degrees

3. A unit that measures the rate at which electric energy is used is a

    (1) volt   (2) watt   (3) calorie   (4) degree

4. Which is *not* a fossil fuel?

(1) uranium   (2) coal   (3) oil   (4) natural gas

5. Plastics, fertilizers, and synthetic fabrics are products commonly made from

(1) coal   (2) oil   (3) gasoline   (4) uranium

6. Fossil fuels were formed from

(1) rocks and minerals

(2) uranium deposits

(3) remains of dead plants and animals

(4) moving water

Base your answers to questions 7–10 on the graphs below, which compare energy used in the United States with energy used in Ohio and New York.

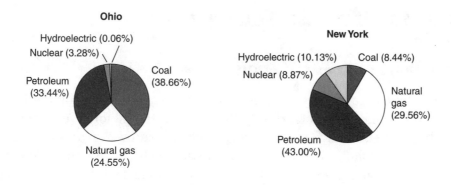

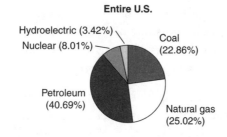

7. Approximately what fraction of Ohio's energy comes from petroleum?

(1) one-half   (2) one-quarter   (3) one-third   (4) two-thirds

8. Compared to the entire United States, Ohio uses

(1) a larger percentage of oil and a larger percentage of coal

(2) a smaller percentage of oil and a larger percentage of coal

(3) a smaller percentage of oil and a smaller percentage of coal

(4) a larger percentage of oil and a smaller percentage of coal

9. To the nearest whole number, what percentage of Ohio's energy comes from fossil fuel?

(1) 33%   (2) 72%   (3) 63%   (4) 97%

10. From the graphs of New York and Ohio, we could reasonably infer that

   (1) Ohio has more nuclear power plants than New York

   (2) New York has more available sources of moving water than Ohio

   (3) coal is more available in New York than in Ohio

   (4) People in Ohio use their cars more than New Yorkers do

Thinking and Analyzing

11. Higher gasoline prices would most likely result from a shortage of

   (1) coal   (2) oil   (3) natural gas   (4) heating fuel

12. What are three ways that you can conserve energy in your home or in your school?

Problems with Energy Sources

Problems with Fossil Fuels

In the United States, most of the demand for energy is met by the fossil fuels: oil, coal, and natural gas. In 2000 and 2001, California experienced shortages of electricity. The shortages were caused by a sharp increase in the price of fossil fuels needed to generate the electricity. Because state law did not allow an increase in cost to the public, some companies found that they could no longer afford to produce electricity. This led to a decrease in the supply of energy available in California.

The burning of fossil fuels creates air pollution. When fossil fuels burn, chemicals that pose dangers to living things and the environment are released into the air. This is especially true of coal.

1. *Coal.* The supply of coal found in the United States is much greater than that of oil or natural gas. Because of its abundance, coal is relatively inexpensive. However, there are serious environmental and health problems involved with its use. The burning of coal contributes greatly to air pollution. Smoke from coal-burning power plants is the main cause of *acid rain*, which is harmful to the ecology of lakes and forests.

   In addition, coal mining is dangerous for people who work in the mines. Breathing air that contains coal dust is unhealthful and can lead to *black lung disease.* Certain coal-mining techniques are also damaging to the environment. Sometimes large areas of land are dug up to reach the coal. This practice, called strip *mining*, destroys topsoil and scars the landscape.

   Mining companies are now required by law to restore the land they have damaged. Advances in technology have made coal mining safer and reduced the amount of pollution caused by burning coal. Nevertheless, these measures have only begun to solve the problems with using coal as a fuel.

2. *Oil and Natural Gas.* Although oil and natural gas produce less air pollutants than coal, they cause other environmental problems. Offshore drilling for oil and transporting oil by ship can lead to accidental oil spills that kill marine wildlife and cause severe pollution of land and sea. Pipelines built to transport oil and gas over land may alter the ecology of areas they cross.

3. *Global Warming.* The burning of any fossil fuel produces carbon dioxide. Carbon dioxide traps heat in Earth's atmosphere much as the glass of a greenhouse traps heat. Some scientists fear that the buildup of carbon dioxide in the atmosphere caused by using fossil fuels may produce a *greenhouse effect*, leading to a warming of Earth's climate. Such global warming could have many harmful consequences for life on Earth.

Problems with Hydroelectric and Nuclear Energy

The production of electricity using moving water or nuclear reactors is generally much "cleaner" (produces less pollution) than energy production with fossil fuels. This is because hydroelectric and nuclear power plants do not involve burning anything. However, even these "clean" energy sources have environmental costs.

1. *Hydroelectric energy.* Building a dam on a river to produce hydroelectric power changes the surrounding area, as you can see in Figure 12-5. The area upriver (behind the dam) is flooded, creating a large lake over land that may have once provided a habitat for wildlife or been used for farming. The area downriver from the dam receives a diminished flow of water. These changes greatly affect the ecology of the area around a dam.

2. *Nuclear energy.* Although nuclear power plants do not cause air pollution, they use water from nearby lakes or rivers to cool their reactors. After cooling the reactor, the water is returned to the environment several degrees warmer than it was. This increase in the

Figure 12-5. *The effect of a hydroelectric dam on the environment.*

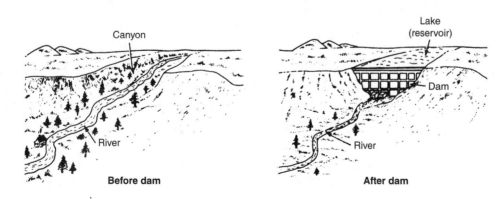

temperature of the water, called ***thermal pollution,*** can be harmful to organisms living in the water.

An even more serious problem with nuclear power is how to dispose safely of the used-up uranium material fuel, known as ***nuclear waste.*** This poisonous, radioactive material must be stored where it will never leak into the environment. Most people do not want nuclear waste stored in, or even transported through, their communities. Disposal of nuclear waste is a difficult problem, and scientists continue to disagree on whether any of the proposed solutions are adequate.

 Review Questions

Multiple Choice

13. Most of the energy consumed in the United States comes from
    (1) nuclear energy   (3) moving water
    (2) fossil fuels   (4) solar energy

14. Which fossil fuel is most abundant in the United States?
    (1) oil   (2) coal   (3) natural gas   (4) uranium

15. Killing of marine wildlife and pollution of shorelines result from accidents involved in
    (1) coal mining   (3) building dams
    (2) storing nuclear waste   (4) transporting oil by ship

16. Acid rain is mainly caused by
    (1) burning natural gas   (3) drilling for oil
    (2) nuclear reactors   (4) burning coal

17. The threat of global warming is mainly associated with
    (1) radioactive wastes
    (2) use of hydroelectric energy
    (3) burning of fossil fuels
    (4) thermal pollution of lakes and rivers

18. Which energy source produces the most air pollution?
    (1) coal   (2) uranium   (3) moving water   (4) natural gas

19. Thermal pollution from nuclear power plants involves
    (1) release of toxic chemicals into the environment
    (2) storage of nuclear wastes
    (3) an increase in temperature of the environment
    (4) a decrease in temperature of the environment

20. The safe storage of hazardous wastes is a problem involved mainly with

   (1) transporting oil

   (2) strip mining

   (3) hydroelectric power

   (4) nuclear power

21. A problem with dams and pipelines is that they

   (1) contribute to air pollution

   (2) produce toxic wastes

   (3) cause thermal pollution

   (4) alter the ecology of surrounding areas

22. The energy source being used to make electricity in the diagram below is

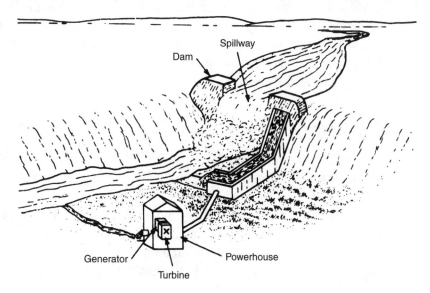

   (1) nuclear power   (2) moving water   (3) a fossil fuel   (4) wind

Thinking and Analyzing

Base your answers to questions 23 and 24 on the reading passage below.

Gasoline-burning vehicles produce exhaust that contains many toxic substances. The pollutants damage the environment, and may be responsible for the rise in the number of cases of asthma and other respiratory problems. Modern technology has developed cars and buses called NGVs (Natural Gas Vehicles) that burn compressed natural gas (CNG). Because natural gas burns cleanly, the exhaust from these vehicles contains mostly carbon dioxide (CO_2) and water (H_2O). Other carbon pollutants have been reduced about 89%.

   Although CNG-powered vehicles are better for the environment than are gasoline-powered vehicles, some environmentalists are urging the use of electric vehicles (EVs) that produce no exhaust at all. They argue that although using CNG will improve air quality, it will not completely solve the problem of global warming.

23. Why is burning CNG preferable to burning gasoline?

24. Why do some scientists prefer battery-powered vehicles to those burning CNG?

Energy for the Future

Energy Conservation

Most energy resources are limited. To guarantee an adequate supply of energy for the future, we must practice conservation. **Conservation** is the saving of natural resources through wise use. This means using resources more efficiently and eliminating unnecessary waste. The methods used in conserving our resources are often summarized as the three R's of conservation—*reduce, reuse,* and *recycle.*

Reducing Energy Consumption

1. *High-efficiency appliances.* We can contribute to conservation efforts by purchasing high-efficiency appliances. These appliances consume less energy than do less efficient appliances while doing the same job. For instance, a car that can travel 48 kilometers on a gallon of gasoline is more efficient than a car that gets only 24 kilometers per gallon. An air conditioner with a high "energy efficiency rating" uses less electricity than one with a low rating, but it cools a room just as well.

   Although a high-efficiency appliance may cost more to buy, it costs less to use. This means that it will save money in the long run, while helping to conserve energy. Many appliances carry an "energy efficiency rating" label that shows how they compare with other models. Figure 12-6 shows an example of such a label.

   There are some disadvantages of high-efficiency appliances. They often require the use of more expensive materials. Many people will not buy more expensive appliances unless they can see that they will save money in the long run. Scientists and engineers are constantly working to develop machines that combine high efficiency with low cost.

   Sometimes, increasing the energy efficiency of a machine affects its performance. Lighter-weight automobiles burn less gasoline than do heavier ones, but may not be as safe in the event of a collision. Cars with larger engines, which can produce greater acceleration, usually burn more gasoline than do cars with smaller engines. Yet many people prefer big, fast cars despite their greater cost and possible harm to the environment.

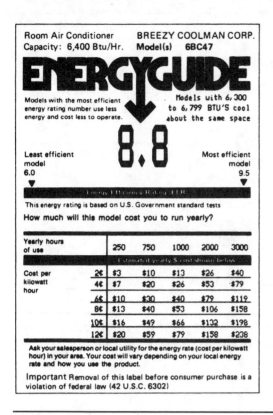

Room Air Conditioner | BREEZY COOLMAN CORP.
Capacity: 6,400 Btu/Hr. | **Model(s) 6BC47**

ENERGYGUIDE

Models with the most efficient energy rating number use less energy and cost less to operate.

Models with 6,300 to 6,799 BTU'S cool about the same space

8.8

Least efficient model
6.0 ▼

Most efficient model
9.5 ▼

Energy Efficiency Rating (EER)

This energy rating is based on U.S. Government standard tests

How much will this model cost you to run yearly?

Yearly hours of use		250	750	1000	2000	3000
		Estimated yearly $ cost shown below				
Cost per kilowatt hour	2¢	$3	$10	$13	$26	$40
	4¢	$7	$20	$26	$53	$79
	6¢	$10	$30	$40	$79	$119
	8¢	$13	$40	$53	$106	$158
	10¢	$16	$49	$66	$132	$198
	12¢	$20	$59	$79	$158	$238

Ask your salesperson or local utility for the energy rate (cost per kilowatt hour) in your area. Your cost will vary depending on your local energy rate and how you use the product.

Important Removal of this label before consumer purchase is a violation of federal law (42 U.S.C. 6302)

Figure 12-6. Example of an appliance's "energy efficiency rating" label.

2. *Insulated buildings.* Energy use at home and at work can be reduced through improved **insulation**. A well-insulated building prevents heat loss in winter and keeps heat out in summer. These benefits can be achieved by constructing walls in two layers, with insulating material in between. Cracks around doors and windows can be sealed with weather stripping for further insulation. With these improvements, less energy is needed to maintain a comfortable indoor temperature year-round.

There are some disadvantages to insulation as well. Some materials that have been used as insulators emit toxic fumes when burned. Safe, efficient insulating materials can be expensive. In some cases, insulation can affect the performance of an appliance. For example, better insulation can greatly increase the efficiency of a refrigerator, but the space occupied by the insulating material may decrease the amount of food the refrigerator can hold.

Reusing Materials

Every time you go to the supermarket you receive one or more bags in which to carry your purchases. Many of these bags could be used more than once, yet we usually throw them away. Today, in an effort to save energy and improve the environment, some supermarkets are offering a slight discount to shoppers who bring their own bags. Reusing resources is efficient

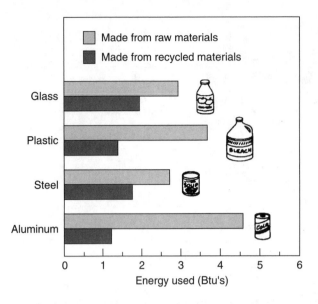

Figure 12-7. *Less energy is consumed when products are made from recycled materials than when they are made from raw materials.*

in terms of energy and waste management. Other reusable products include glass and plastic bottles, cans, and even worn clothing. Some disadvantages of reusing material are the time and effort required for saving and collecting them.

Recycling

This practice also helps conserve energy resources. The graph in Figure 12-7 shows that making bottles and cans from recycled materials consumes less energy than does making those products from raw materials.

As you can see in the figure, less energy is used to make glass containers from recycled materials than from raw materials. However, the collection of glass for recycling requires the cooperation of large numbers of people, who must separate glass containers from other waste. These must be cleaned, collected, and transported to a recycling plant. Sometimes, the cost of collection and transportation is so high that the process no longer saves money. However, recycling is important even when money is not the issue. When materials such as glass and paper are recycled, they no longer have to be disposed of. Recycling benefits the environment by reducing the amount of waste that must be dumped in landfills or burned in incinerators.

It's Your Decision

In many places in the United States, recycling is the law. People are required to sort their trash so that recyclable materials can be collected. This requires extra time and effort, but they have decided, through their elected representatives, that the time and effort are worth it. You will be making decisions

Table 12-2. *The Cases For and Against Recycling Paper*

For	Against
Preserves forests, which soak up carbon dioxide and provide habitat for wildlife	Trees can be replaced by new trees, minimizing damage to forests; may cause job layoffs in timber industry
Decreases amount of solid waste in landfills	Expensive
Decreases air pollution in areas where paper is burned in incinerators	Separation of used paper from other waste is time-consuming

like these in the future. Let us examine some of the issues you should consider in making your decisions.

Have you ever noticed that a greeting card printed on recycled paper costs more than one printed on new paper? Why should we recycle paper when it saves us neither energy nor money? Paper is made from wood from our forests. By using recycled paper, we are cutting fewer trees, which benefits the environment. Table 12-2 lists some arguments for and against the recycling of paper.

The decision of whether to recycle affects people's jobs and businesses as well as the environment. Environmentalists and businesspeople will try to convince you of their points of view. Listen carefully to both sides before you make your decisions.

Renewable and Nonrenewable Resources

Earth's supply of fossil fuels is being used up rapidly. We continually remove these resources from the earth, but we cannot replace them. Nature does not create new deposits of oil, coal, and natural gas within the time span of human history. For this reason, fossil fuels are considered *nonrenewable resources.* Uranium is also a nonrenewable resource.

Renewable resources are those that can be replenished by nature within a relatively short time span. Moving water, wind, plants, and sunshine do not run out as we use them. Natural processes are constantly replacing them. Table 12-3 lists some renewable and nonrenewable energy resources.

Table 12-3. *Energy Resources*

Renewable	Nonrenewable
Hydroelectric	Oil
Solar	Coal
Wood	Natural gas
Wind	Nuclear

Using Renewable Resources

Even if we practice conservation, our supply of nonrenewable energy resources may not be sufficient to meet the energy demands of the future. Renewable resources offer alternatives to fossil fuels and radioactive minerals. Unlike nonrenewable resources such as oil, coal, and uranium, renewable energy resources cannot run out. They also cause fewer environmental problems than fossil fuels and nuclear energy do. For these reasons, scientists and engineers are seeking more and better ways to use renewable resources for our growing energy needs.

As you know, moving water can be used to run generators and produce electricity. The natural water cycle of evaporation, condensation, and precipitation renews the water supply that feeds the rivers used for this purpose. However, not all areas have rivers suitable for producing hydroelectric energy.

The wind can be used to generate electricity by turning the blades of a *wind turbine* (see Figure 12-8). In windy areas, wind turbines can provide safe, clean electricity. But the wind is not as constant and reliable as a flowing river. When there is only a slight wind or the air is calm, little or no electricity is produced.

Plant matter and animal wastes can be burned to produce heat, or they can be changed to other fuels. For instance, decaying plant matter and animal wastes produce *methane,* the main component of natural gas. Methane produced in this way is a renewable resource, unlike natural gas found underground. At present, however, converting these materials into fuel on a large scale is too expensive to be practical. Unfortunately, *every* energy source has both advantages and disadvantages, as outlined in Table 12-4.

Solar Energy

The primary source of energy on Earth is the sun. Energy from the sun is called **solar energy**. The energy in fossil fuels came originally from sunlight. This light was absorbed by plants during photosynthesis million of years ago. The moving water used for hydroelectric energy is replenished by the water cycle, which is powered by the sun's energy. Wind, which can be used to make electricity, is caused by the sun's heating of the atmosphere.

Figure 12-8. Wind turbines use the renewable energy of the wind to produce electricity.

Table 12-4. Advantages and Disadvantages of Energy Sources

Energy Source	Advantages	Disadvantages
Oil	Efficient; can be converted into different types of fuel	Causes air pollution; risk of spills while drilling or transporting; limited reserves in U.S.; nonrenewable
Natural gas	Available in U.S.; clean	Difficult to store and transport; mostly nonrenewable
Coal	Abundant in U.S.; inexpensive	Causes air pollution and acid rain; mining practices may be harmful to miners' health and destructive to the environment
Nuclear	Abundant fuel in U.S.; does not cause air pollution; can meet long-term energy needs	Causes thermal pollution; creates radioactive waste; risk of accidents releasing radioactivity into environment; uranium mining harmful to miners' health
Hydroelectric	Does not cause air pollution; inexpensive; renewable	Not available in all areas; affects local ecology
Wind	Does not cause pollution; clean, inexpensive, renewable	Not practical for large-scale generation; not always reliable (wind is not constant)
Solar	Does not cause pollution; clean; renewable	Expensive to convert into usable form; not always reliable (depends on the weather)
Plant matter and animal wastes.	Renewable	Expensive to convert into usable form; inefficient

People have found ways to use the sun's energy directly to provide heat and hot water for homes, offices, and factories. For example, a device called a *solar collector* absorbs solar energy and converts it into heat energy. The heat is transferred to water circulating through the collector. This hot water can be used to run a home heating system, as shown in Figure 12-9.

Figure 12-9. Example of capturing and using solar energy to run a home heating system.

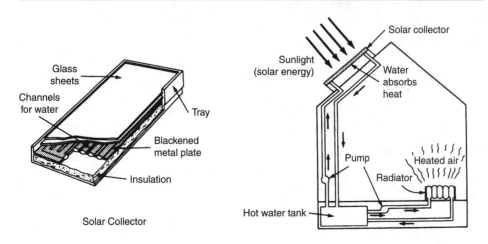

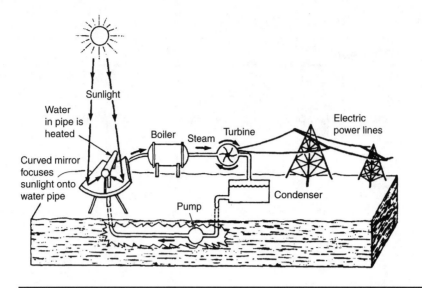

Figure 12-10. *Solar energy changes water into steam, which turns turbines to produce electricity.*

People have also developed ways to transform solar energy into electric energy. For instance, a *solar cell* is a device that converts light directly into electricity. Some calculators and light meters in cameras use solar cells instead of batteries for energy. However, to generate large amounts of electricity this way requires a huge number of these cells, which is very expensive.

Heating water with solar energy can also produce electricity. Water is heated to a boil by using mirrors to focus and concentrate sunlight (see Figure 12-10). The boiling water changes into steam, which turns turbines to produce electricity. This method is more economical than using solar cells to make electricity, but it is still more expensive than using fossil fuels. However, as fossil fuels become more scarce, this situation may change.

 Review Questions

Multiple Choice

25. Energy resources can be conserved by all of the following means *except*

(1) better insulation

(2) using high-efficiency appliances

(3) increased mining of coal

(4) increased use of solar energy

26. Which is a renewable energy resource?

(1) moving water   (2) uranium   (3) coal   (4) oil

27. The primary source of most energy on Earth is

(1) moving water   (2) the sun   (3) coal   (4) wind

28. Because natural gas taken from underground is not quickly replaced by nature, it is considered a

(1) renewable resource   (3) pollutant

(2) nonrenewable resource   (4) solar energy source

Thinking and Analyzing

Base your answer to question 29 on the passage below:

Gasohol is a combination of gasoline and alcohol. Gasoline comes from oil, and alcohol is made from plant matter. Gasohol was developed to decrease our use of oil. However, gasohol costs more than ordinary gasoline and is therefore not commonly used.

29. This paragraph suggests that

(1) gasohol will soon replace gasoline

(2) gasohol causes less pollution than gasoline

(3) renewable resources are less efficient than nonrenewable resources

(4) alternative energy sources may have both advantages and disadvantages

30. In the diagram below,

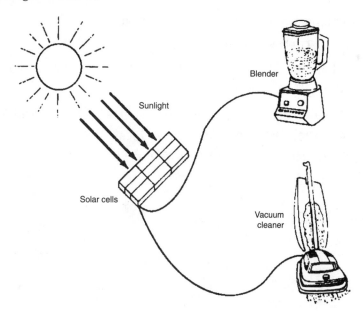

(1) electric energy is being changed into solar energy

(2) solar energy is being changed into electric energy

(3) chemical energy is being changed into solar energy

(4) solar energy is being changed into chemical energy

31. Which are arguments for recycling paper?

   A. Burning paper causes air pollution.

   B. Paper comes from trees, which are a renewable resource

   C. Collection of paper for recycling is time consuming.

   D. Used paper adds to the waste that is stored in landfills.

   (1) A only   (3) A, B, and C only

   (2) A and D only   (4) A and B only

Questions 32–34 are based on the graph below, which shows the monthly gas and electric bills of a family for one year.

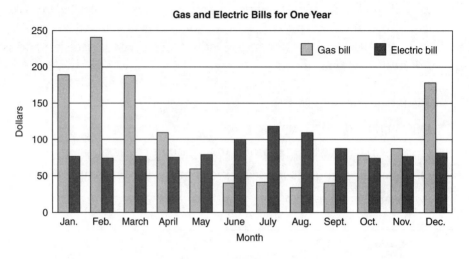

Gas and Electric Bills for One Year

32. These bills are probably for a home that is heated by

   (1) oil   (2) electricity   (3) wood   (4) gas

33. For which month was the electric bill highest? Why do you think that month's bill is highest?

34. For which month were the gas and electric bills most nearly the same?

35. What is a renewable resource? How does it differ from a nonrenewable resource? What are two examples of each type?

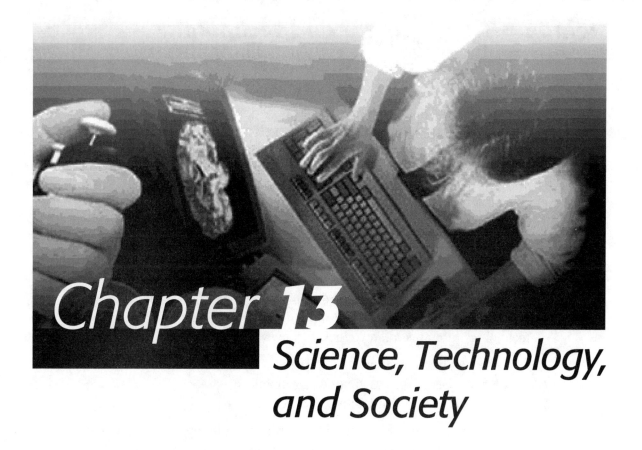

Chapter *13*
Science, Technology, and Society

Points to Remember

▷ The emphasis in science is on gaining knowledge of the natural world by asking questions; the emphasis in technology is on finding practical ways to apply that knowledge to solve problems. Science and technology frequently help to advance each other. The search for knowledge is a driving force for science and technology.

▷ We all interact with the products of technology in almost everything we do.

▷ Technology is used to extend or improve our abilities, and many products of technology affect the environment in some way.

▷ Science, technology, and society are constantly interacting. Often, a change in one of these areas will affect the other two.

▷ Every technological process or device has advantages and disadvantages associated with its use.

Relationship of Science and Technology

Science and Technology

Science and technology affect the lives of people all over the world. *Science* is the process of asking questions and seeking answers to gain an understanding of the natural world. By providing insight into the workings of nature, science helps us predict the outcome of physical events.

Some questions that science attempts to answer include:

> What is the nature of matter?
> How did the universe come into being?
> How did life evolve on Earth?

Technology uses scientific knowledge and other resources to develop new products and processes. These products and processes help solve the problems and meet the needs of individuals or society. Some problems that technology attempts to solve include how to:

- increase the gas mileage of cars
- increase the productivity of farmland
- control industrial pollution

While the emphasis in science is on gaining knowledge of the natural world, the emphasis in technology is on finding practical ways to apply that knowledge to solve problems. There are three major fields of science: life science, earth science, and physical science. Each of these fields contains a number of more specific sciences (see Figure 13-1). Biologists, chemists, and geologists are some types of scientists. Engineers, computer programmers, and medical technicians are examples of workers in the fields of technology.

Science and Technology Advance Each Other

Science and technology frequently help to advance each other. Scientific discoveries often lead to the development of new or better devices and processes. These technologies may, in turn, lead to new discoveries or to a better

Figure 13-1. *The major fields of science.*

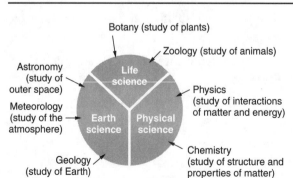

Table 13-1. Relation of Scientific Principles and Technology

Scientific Principle	Technological Device/Process
Low temperatures kill or reduce growth of microorganisms	Refrigerators and freezers
Sunlight contains energy	Solar heating systems and solar cells
Splitting atoms of radioactive elements produces heat	Nuclear power plants
Every action produces an equal and opposite reaction	Rocket engines and jet engines

understanding of scientific principles. For instance, scientists discovered various properties of light, such as how light is bent when it passes through different types of lenses. This knowledge led to the invention of the telescope and the microscope. Using these devices, scientists have made many more discoveries about the natural world. Every technological advance is based in some way on scientific principles, as the examples in Table 13-1 suggest.

In fact, much technology involves knowledge from more than one field of science. For example, the artificial heart shown in Figure 13-2 involves knowledge from both the life sciences (the structure of the human heart) and the physical science (the mechanical principles of how the heart works).

Search for Knowledge

Many people throughout the ages have had a desire to learn about the universe and the world around them. This search for knowledge is the essence of science and technology. It drives scientists to find answers to many questions regarding life, such as, who are we and where are we in this vast universe? It also provides the incentive to develop newer technologies to answer the questions. For example, improved and bigger telescopes help us

Figure 13-2. Scientific knowledge made possible the development of the artificial heart.

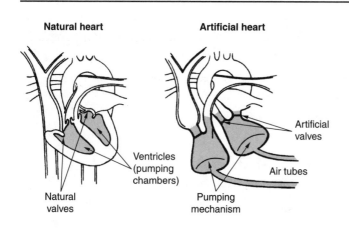

to see further into the universe. Table 13-2 below demonstrates how space exploration provides a good example of science and technology assisting each other in the search for knowledge about the universe.

Technology as a System

Some technological processes and devices can be thought of as systems. A *system* is a group of related elements or parts that work together for a common purpose. The parts of a system act in a series of steps. These steps are input, comparison and control, process, output, and feedback.

A home heating system with a furnace and a thermostat can be viewed in terms of these steps (see Figure 13-3). Setting the thermostat to the desired temperature is the input. The thermostat compares the actual room

Table 13-2. *Space Exploration Highlights for Years 1957–2001*

1957	U.S.S.R. places Sputnik 1 into orbit around Earth.
1959	Satellites orbit the sun and moon.
1960	First weather satellite placed in orbit.
1961	First Astronaut and Cosmonaut sent into space.
1964	Ranger 7 takes close-up pictures of the moon.
1965	First Cosmonaut and Astronaut make space walk.
	Close-up images of the moon and Mars.
1966	First spacecraft to orbit moon, and first to land softly on the moon.
1967	Data received from capsule dropped to Venus's surface.
1968	First manned spacecraft to orbit the moon and return.
1969	First landing of astronauts on the moon.
	High-resolution pictures of Mars.
1970	First probe to land softly on Venus.
1971	Space station is placed in orbit.
1972	First close-up pictures of Jupiter.
1973	Skylab placed in orbit and maintained by crews.
1974	Mariner 10 takes close-up images of Mercury.
1977–89	Voyager spacecrafts fly by and gather information from Jupiter, Saturn, Neptune, and Uranus.
1986	Spacecraft gathers information from Halley's Comet.
1989	Spacecraft crashes into Jupiter and sends information back to Earth.
1990	Spacecraft placed in orbit around the sun.
	Hubble Telescope placed in orbit around Earth.
1996	Spacecraft placed in orbit around Mars.
1998	Russia launches first mission to construct International Space Station.
1999	Launched spacecraft to gather samples of comet and return to Earth.
2001	Spacecraft studies Eros, a near-Earth asteroid.

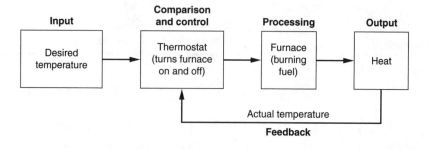

Figure 13-3. *A home-heating system is a technological example of a feedback system.*

temperature to the set temperature and controls the furnace, turning it on if the room's temperature is too low. The burning of fuel in the furnace is the process that produces the output—heat. The changing room temperature provides feedback to the thermostat, which turns the furnace off when the desired temperature is reached. In this way, the system maintains a constant indoor temperature.

Technology in Use

Everyone interacts with the products of technology. In almost everything you do, when you wear clothing, sleep on a bed, watch television, eat with a knife and fork, ride a school bus, you interact with the products of technology. People use technology for a number of reasons (see Figure 13-4). To extend or improve our abilities, we use radios and telephones, calculators and computers, binoculars and telescopes, and other devices. Machines and appliances help us do work that requires more than human strength, and at faster speeds than are humanly possible. To overcome physical disabilities, people use devices like eyeglasses, hearing aids, and heart pacemakers.

Many products of technology are used to change our environment. For example, we use electric lights so we can continue our activities after nightfall. Every technological process or device affects the environment in some

Figure 13-4. *Some technological devices are used to extend human abilities or to overcome disabilities.*

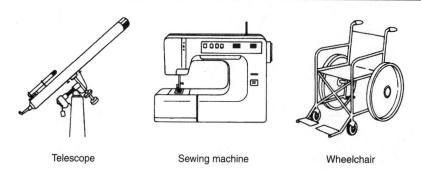

Telescope   Sewing machine   Wheelchair

Table 13-3. *Technology and Our Environment*

	Technology	How It Affects the Environment
Beneficial Effects	Electric Lights	Extend light into night time
	Dams	Store water, generate electricity, and create lakes for recreation
	Sewage-treatment plants	Reduce pollution of rivers and lakes
Harmful Effects	Electric generators	Consume natural resources and cause pollution
	Cars, boats, and airplanes	Consume natural resources and cause pollution

way. Some of these effects may be harmful. Lightbulbs and many other appliances require electricity. The production of electricity may use precious natural resources or cause pollution of air and water. However, technology can also be used to protect the environment, as with sewage-treatment plants and pollution-control devices in cars and factories. Table 13-3 gives some examples of beneficial and harmful effects of technology on the environment.

 **Laboratory Skill**

Graphing Data

The table below lists the numbers of different word-producing machines used in a newsroom from 1955 to 2000, at five-year intervals. These numbers show how the usage of manual typewriters, electric typewriters, and word processors has changed over time.

Numbers of Different Word-Producing Machines in a Newsroom, 1955–2000

Year	Manual Typewriters	Electric Typewriters	Word Processors
1955	45	0	—
1960	35	10	—
1965	25	20	—
1970	10	40	—
1975	0	50	0
1980	—	40	10
1985	—	20	40
1990	—	5	50
1995	—	5	52
2000	—	1	55

(Continued)

The relationships among the three sets of data may not be immediately clear from the table. However, if the same data are presented in a graph, these relationships become much easier to interpret.

The graph below shows only the data for usage of manual typewriters (represented by triangles). Copy the graph on a separate sheet of graph paper. Then enter the data for electric typewriters, using circles, and the data for word processors, using squares. (It may be helpful to use a different color pencil for each set of data.) Then answer the following questions.

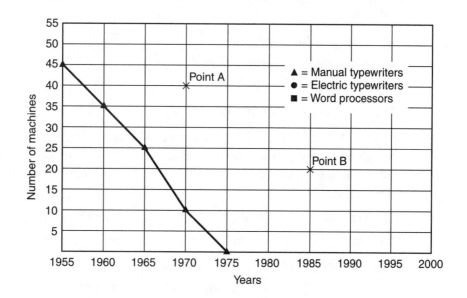

QUESTIONS

1. Point A on the graph represents the number of
   (1) manual typewriters in use in 1970
   (2) electric typewriters in use in 1980
   (3) electric typewriters in use in 1970
   (4) word processors in use in 1970

2. Point B represents the number of
   (1) electric typewriters in use in 1985
   (2) word processors in use in 1980
   (3) word processors in use in 1986
   (4) none of the above

3. If the trend for word processors continued, the number of word processors in use in 2005 would have been
   (1) less than 2000
   (2) more than 2000
   (3) the same as 2000
   (4) impossible to determine

Review Questions

Multiple Choice

1. Using scientific knowledge to develop new products or processes is called

   (1) science   (3) technology

   (2) industry   (4) renewing resources

2. Using scientific knowledge about magnetism and electricity to build an electromagnet is

   (1) a scientific discovery

   (2) a technological development

   (3) predicting future physical events

   (4) observing the natural world

3. Using scientific knowledge, engineers built a space probe and sent it to the planet Jupiter. The probe sends data about Jupiter back to Earth, adding to our scientific knowledge. Which statement does this best demonstrate?

   (1) New technology sometimes builds on past technology.

   (2) Advances in technology cause some devices to become obsolete.

   (3) Technology affects our environment.

   (4) Science and technology help to advance each other.

4. Which of these technological devices affects the environment in some way?

   (1) coffee pot   (3) air conditioner

   (2) washing machine   (4) all of these

5. Which of the following statements is most correct?

   (1) Few people interact with the products of technology.

   (2) Only people in the United States interact with the products of technology.

   (3) Only adults interact with the products of technology.

   (4) Everyone interacts with the products of technology.

6. In this air-conditioning system, the output is the

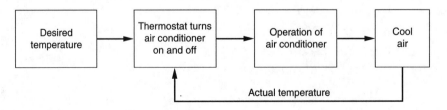

(1) desired temperature   (3) actual temperature

(2) cool air   (4) thermostat

7. The difference between the two heating systems shown below is that system A is lacking

    (1) input

    (2) output

    (3) feedback

    (4) processing

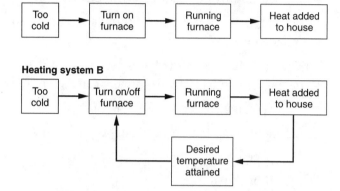

8. A technological device or process may become outdated, or obsolete, when a new device or process is developed that does its job better. For example, the slide rule to do mathematical calculations has been replaced by the hand calculator. Another example of an older device that has become obsolete and the new device that replaced it is

    (1) buses and cars

    (2) fountain pens and ball point pens

    (3) eye glasses and contact lenses

    (4) shoes and sneakers

Thinking and Analyzing

9. The chart below lists three general uses of technology. Examples of uses 1 and 2 are given in the chart. Which three examples would best fit in the third column?

Chart: Uses of Technology

Affect Our Environment	Overcome Disabilities	Extend Our Abilities
Air conditioner	Hearing aid	
Space heater	Wheelchair	
Dam on a river	Eyeglasses	

    (1) heart pacemaker, furnace, traffic light

    (2) binoculars, telephone, calculator

    (3) hammer, furnace, videocassette recorder

    (4) airplane, plumbing, coal mining

10. Space exploration began in the 1960s and 1970s. It progressed from simple to complex, for example 1) orbit Earth, 2) orbit Moon, 3) orbit Mars, etc. The most likely reason for this progression was

    (1) the invention of rockets

    (2) the increase in science knowledge

    (3) the improvements in technology

    (4) the increase in science knowledge and improvements in technology

11. The search for knowledge is an important issue for many people, especially scientists. An example of the search for knowledge is

    (1) exploring the bottom of the ocean

    (2) identifying the components of atoms

    (3) space exploration

    (4) all of the above

Interaction of Science, Technology, and Society

Effects of Science and Technology on Society

Science, technology, and society are constantly interacting with one another (see Figure 13-5). Often a change in one area will affect the other two. For example, scientific discoveries about the structure of matter led to many technological developments, including the production of microprocessors on tiny silicon chips. These "microchips" made possible many new products that have affected society by improving health care, communications, and transportation.

Figure 13-5. Science, technology, and society are constantly interacting and affecting one another.

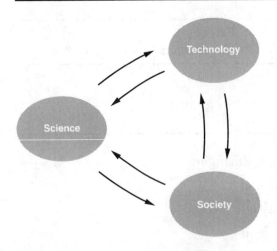

Table 13-4. Ways Science and Technology Have Affected Society

New or Improved Products Raised Standard of Living

Health care products allow us to live longer, healthier lives.

Work-saving home appliances provide more leisure time.

Created New Businesses and Industries

Film processing developed into motion picture industry.

Music recordings changed from phonograph records to tapes to CDs.

Eliminated Businesses and Industries

Refrigerators and freezers eliminated the need for the ice-cutting industry.

Home heating oil eliminated need for coal companies providing home delivery.

Career Choices and Job Opportunities Eliminated, Modified, and Created

Telephone operators replaced by computers.

Typists had to learn to use computers for word processing.

Computer systems needed computer analysts for implementation.

Solved Society's Problems

Vaccines eliminated certain diseases.

Communications systems, such as cell phones and the Internet, allow people separated by great distances to work together.

Our culture, economy, and social systems are often affected by developments in science and technology. During the 1800s, the United States was transformed from a mainly agricultural society to a highly industrialized society. This period of cultural, economic, and social change was caused by the development of industrial machinery and new ways to power it. Other ways that science and technology have affected society are shown in Table 13-4.

While science and technology have solved many problems, they have also created problems. Pollution of the environment and disposal of garbage and hazardous waste are problems caused, in part, by science and technology (see Figure 13-6). Solving such problems requires the help of people working in government, industry, science, and technology.

Figure 13-6. *Disposal of garbage is a social problem. Much of the garbage consists of products of science and technology.*

Effects of Society on Science and Technology

Society also affects science and technology in many ways. New technology is often developed in response to the needs of individuals or society. For example, the need to help people overcome diseases and disabilities has encouraged the development of new medical procedures, such as chemotherapy and laser surgery, and new devices, like artificial organs and limbs.

The attitudes of people in a society may influence the direction of scientific research and technological development. In our society, public opinion has encouraged research to find a cure for AIDS. In contrast, public attitudes have largely discouraged the use of animals to test the safety of new cosmetic products.

Acceptance and use of an existing technology can also depend on people's attitudes. An example is nuclear energy. Most people agree that nuclear energy has both benefits and drawbacks. However, people disagree about whether or not its benefits outweigh its dangers. Public attitudes against nuclear energy have led some countries to ban its use. Yet, other countries generate most of their electricity with nuclear energy. Public opinion will undoubtedly influence the future of nuclear energy (see Figure 13-7).

Historical, Political, and Social Factors Affect Science

Historical, political, and social factors have always influenced science. These factors have caused science at times to surge forward, at other times to be slowed, and even sometimes to become a controversial social issue. Some examples will demonstrate how science can be affected by each of these factors.

Figure 13-7. People's attitudes toward nuclear power will affect the acceptance and use of this technology.

Historical Factor—Galileo Galilei (1564–1642), an Italian scientist, built one of the first telescopes. With it he studied astronomical objects such as the sun, moon, and stars. He recorded his observations and made numerous calculations of objects' positions and motions. Among many conclusions, Galileo supported an unpopular theory that the sun was stationary and located in the center of the solar system (heliocentric theory). He also stated that Earth traveled around the sun. However, the church at that time did not support the heliocentric theory. It condemned Galileo, saying that his beliefs were absurd and philosophically false. Galileo was imprisoned in his home for the rest of his life.

The church supported an historical belief that Earth was the center of the universe and that the sun, moon, stars, and other astronomical objects revolved around Earth (geocentric theory). Galileo's evidence was unacceptable because of the historical beliefs of the church.

Political Factor—On October 4, 1957, the Soviet Union placed a satellite called *Sputnik 1* into orbit around Earth. The Soviet Union was jubilant over its success in placing the first artificial satellite in orbit. This event caused alarm within the United States. Politicians saw this achievement as a potential threat to our military defense and a sign of decline in our science capabilities. In January 1958, the United States placed a satellite, *Explorer I*, into space, demonstrating that it too was capable of a similar feat. These events marked the beginning of "the space race" between the United States and the Soviet Union.

The space race remained highly competitive for the next three decades. The launching of *Sputnik 1* caused the United States to recognize that in 1957 there was a technological gap between it and the Soviet Union. This realization resulted in a major initiative to improve science education in the United States. The political competitiveness between the two countries did much to promote science and technology in this country. Since the breakup of the Soviet Union, the competition has not been as keen. In fact, the United States and Russia are working together on the International Space Station.

Social Factor—Genetic engineering is the ability to change the genetic makeup of plants and/or animals for the purpose of improving some of their traits. Improving corn and wheat plants genetically to increase their yield would help to eliminate hunger in the world. Using biotechnology to replace disease-producing genes with healthy genes in humans could wipe out certain diseases and increase life expectancy.

However, ethical issues regarding who decides what crops are improved and what diseases are eliminated could produce social unrest. Countries with this knowledge would possess power and control over countries without this ability. In recent years, genetic engineering has become a major controversial issue for those who recognize the pros and cons of its use.

Global Effects of Technology

Technology used in one country may have an international or global impact. For instance, in 1985, an accident at the Chernobyl nuclear power plant in the former Soviet Union released radiation that affected several neighboring countries. The radiation contaminated livestock, crops, and water. Another example is acid rain. Industries in the Midwest sometime create air pollution that drifts eastward with the prevailing winds. This causes acid rain to fall in New York, New England, and parts of Canada.

A possible global effect of technology is the destruction of Earth's ozone layer. Certain chemicals used in refrigerators, air conditioners, and spray cans are destroying the layer of the ozone gas that exists high in the atmosphere. The ozone layer screens out much of the sun's dangerous ultraviolet radiation. The destruction of this layer could cause an increase in skin cancer and may have harmful effects on Earth's ecology.

On the positive side, people around the world interact more frequently because of technological advances. Communication satellites let us make phone calls or send E-mail by computers to people on other continents. The Internet allows us to access information quickly, among other things. We can also view newsworthy events as they happen on television with the help of satellites (see Figure 13-8). Technology helps us predict natural disasters like earthquakes and hurricanes, informs us immediately when they occur, and it provides the means of sending aid to victims of these disasters.

Figure 13-8. *Communication satellites enable us to view distant events on television.*

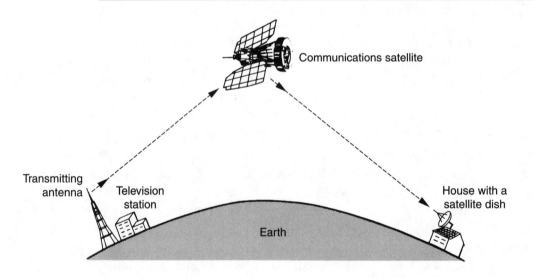

Multiple Choice

12. Society affects technology by

    (1) having problems that need to be solved

    (2) providing funds for research and development

    (3) its attitudes toward new research or products

    (4) all of the above

13. An example of a job that has been created by recent developments in technology is

    (1) farmer

    (2) astronaut

    (3) schoolteacher

    (4) postal worker

14. Developments in microelectronics and computer science are changing the United States from a largely industrial nation into one more dependent on information services. This is an example of

    (1) science helping technology to advance

    (2) society affecting science and technology

    (3) science and technology affecting society

    (4) science and technology solving society's problems

15. Carbon dioxide released into the atmosphere by industry and cars may cause a warming of Earth's climate, called the greenhouse effect. This is an example of

    (1) a problem solved by technology

    (2) people's attitudes affecting the use of technology

    (3) new industries being created by technology

    (4) a global impact of technology

16. In the past, lumberjacks chopped down trees using axes and handsaws. Today, they mostly use motorized chain saws. This example demonstrates that

    (1) technology has created new jobs

    (2) technology has modified some jobs

    (3) technology has eliminated some jobs

    (4) technology has not affected lumberjacks

17. An example of an event that politically increased the need for science knowledge in the United States was

    (1) the election of the U.S. president in 2000

    (2) the landing of a spacecraft on the Moon

    (3) the invention of the television

    (4) the launching of *Sputnik 1* in 1957 by the Soviet Union

Thinking and Analyzing

18. In Europe, many people use irradiated milk (milk subjected to radiation that kills microorganisms). Containers of such milk can be stored at room temperature for a long time if unopened. In the United States, however, some people regard irradiated milk with suspicion, so its use is only slowly becoming accepted.

The above paragraph illustrates that

    (1) technology has affected society by raising our standard of living

    (2) people's attitudes can affect acceptance and use of technological devices or processes

    (3) products of technology may have an international impact

    (4) technology has caused some industries to become obsolete

19. The table below shows how the percentage of the workforce in three fields has changed over the years. From 1800 to 2000, the percent of the workforce in industry has

Percent of Work Force in Three Fields, from 1800 to 2000

Year	Agriculture	Information	Industry
1800	75%	5%	20%
1850	60%	5%	35%
1900	40%	5%	55%
1950	10%	20%	70%
2000	5%	75%	20%

    (1) increased steadily

    (2) decreased steadily

    (3) first increased, then decreased

    (4) first decreased, then increased

20. The following graph shows how average life expectancy has changed over time. This change is most likely a result of

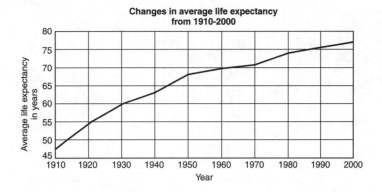

Changes in average life expectancy
from 1910-2000

(1) advances in medical technology

(2) harmful effects of technology on the environment

(3) advances in educational technology

(4) advances in communication technology

21. Which answer is the best example of the relationship shown in the diagram below?

(1) principles of electricity → development of electrical appliances → more leisure time for people

(2) effect of cold on microorganisms → people eat fresher, healthier food → development of refrigerators and freezers

(3) principles of nuclear energy → development of nuclear power plants → development of nuclear weapons

(4) development of radio and television → principles of electromagnetic waves → home entertainment for people

Making Decisions about Technology

Technology Increases Our Choices

People have more choices in their everyday lives because of the products of technology. For example, cable and satellite television, videocassette recorders, and CD and DVD players have increased our choices in home entertainment. Figure 13-9 on page 328 shows some leisure activities that are outgrowths of technology.

Technology has given us more forms of transportation to choose from. For instance, people may travel from New York City to Chicago by car, bus, train, or airplane. Shoppers may choose from a wide selection of home appliances produced by technology. You can even shop "on-line." DVD players, microwave ovens, computers, and graphing calculators are just a few of these products.

| Waterskiing | Bicycle riding | Scuba diving |

Figure 13-9. *Many leisure-time activities are outgrowths of technology.*

Assessing Technology

Every technological process or device has advantages and disadvantages associated with its use, providing both **benefits** and **burdens** for people and the environment. For instance, the automobile has given people greater mobility and contributed to our nation's prosperity. However, cars and trucks contribute to air pollution and lead to deaths and injuries in traffic accidents. Table 13-5 lists benefits and burdens of some technological devices and processes.

Technological processes and devices should be assessed by their advantages and disadvantages. When a device or process is adopted for use, information on its short-term and long-term effects should be continuously collected and evaluated. This helps us to identify and compare the benefits

Table 13-5. *Benefits and Burdens of Technology*

Technological Process or Device	Benefits	Burdens
Nuclear energy	Additional clean electricity	Risk of accidents; radioactive wastes
Pain-killing drugs	Treat diseases, relieve pain	Addiction through abuse
Computers	Increased ability to process data	Loss of jobs; health problems from computer keyboards
Space travel	Increased knowledge	High financial cost
Life-sustaining medical devices	Keep people alive	Decisions about when to use or remove them
Automobile	Increased mobility	Increased pollution; deaths and injuries
Chemical fertilizers	Increased agricultural yields	Upset ecology of lakes and streams; possible harm to humans
Artificial sweeteners	Convenience for diabetics and dieters	Increased risk of cancer
Refrigerants	Storage and preservation of food	Released into atmosphere, it destroys ozone layer
Herbicides and Pesticides	Destroy unwanted plants and insects	Upset natural food chain of some animals; possible harm to humans

and possible adverse consequences of the technology for people and the environment, both for present and future generations.

Our society monitors the effects of many technological devices and processes, including medical treatments, food additives, industrial chemicals, and processes for generating electricity. Various government agencies and public-interest groups perform this task.

Technology and Decision Making

Decisions about the use of technology must be made almost constantly. To make these decisions wisely, both short-term and long-term consequences should be considered. Sometimes the short-term benefits of a technology outweigh its long-term burdens. For example, dentists agree that the benefits of using X rays to find cavities in your teeth outweigh the possible long-term dangers of brief exposure to the radiation (see Figure 13-10).

In other cases, long-term benefits may outweigh short-term burdens. Wearing a seat belt in a car may be a momentary discomfort. Over time, however, the use of seat belts reduces deaths and injuries from car accidents. Society's consideration of short-term and long-term effects has led to using unleaded gasoline for cars (which is less polluting), and the recycling of cans, bottles, and newspapers (to reduce waste).

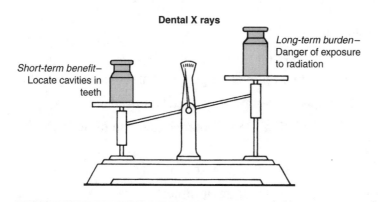

Figure 13-10. *"Weighing the benefits and burdens of technology."*

Review Questions

Thinking and Analyzing

22. Should government regulate the use of technological devices and processes for the good of society? Some social problems can be solved or lessened by laws that regulate technology. But besides providing benefits, such laws may also burden us in some ways. Some laws may infringe on our right to privacy or freedom of choice. Other laws may be

costly to put into action and enforce. Examples of laws that regulate technology are listed in the table below, along with their benefits. Which law has its benefit and burden given in the wrong order?

Problem	Law	Benefit	Burden
Water pollution from industries	Effluent controls law	Cleaner water	Higher industrial costs
Head injuries from motorcycle accidents	Mandatory motorcycle helmet law	Reduces freedom of choice; uncomfortable	Fewer head injuries in motorcycle accidents
Crimes committed with easily obtained handguns	Handgun control law	Less crime; harder for criminals to get guns	Harder for people to get guns for protection or hunting
Bottle and can litter	Bottle deposit bill	Cleaner streets and sidewalks	Inconvenience of returning bottles and cans

(1) effluent controls law

(2) mandatory motorcycle helmet law

(3) handgun control law

(4) bottle deposit bill

23. The development of computer technology and use of the Internet opened up many new career options for people. This is an example of

(1) technology developed in response to society's needs

(2) people's attitudes affecting the acceptance and use of new technology

(3) increased choices brought about by technology

(4) technology affecting the global environment

24. Which answer lists both a benefit and a burden of using the Internet?

(1) provides entertainment, and provides up-to-date news

(2) provides entertainment, and causes higher utility bills

(3) causes eye strain, and discourages reading books

(4) provides useful educational tool, and provides entertainment

25. Most oil tankers have single hulls that break open fairly easily if the ship runs aground, causing oil spills. However, some tanker ships are now being built with double hulls that are more resistant to breaking. This best illustrates that

(1) technology has increased our choices in life

(2) a technology may be modified to reduce or eliminate its disadvantages

(3) government should constantly monitor technology to determine possible adverse consequences

(4) decisions about technology often involve trade-offs between benefits and burdens

26. People with diabetes, who cannot eat sugar, can safely use the artificial sweetener saccharin. It also has fewer calories than sugar and causes less tooth decay. However, some experiments have shown that using saccharin may increase the risk of getting cancer. Saccharin provides

(1) a benefit only

(2) a burden only

(3) both a benefit and a burden

(4) neither a benefit nor a burden

27. In 1984, the bottle deposit bill was passed. This law requires people to pay an extra five cents for every bottle or can of beverage. The deposit is refunded when the empty bottles and cans are returned. The graph indicates that the bottle deposit bill has

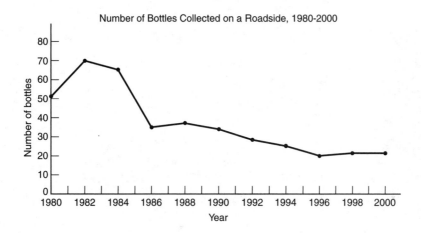

Number of Bottles Collected on a Roadside, 1980-2000

(1) increased roadside litter

(2) decreased roadside litter

(3) had no effect on roadside litter

(4) its effect cannot be determined from the graph

28. Big cars get poor gas mileage, but they are safer than small cars in the event of a traffic accident. Small cars get very good gas mileage, but offer little protection in an accident. For these reasons, many people buy medium-sized cars, which get fairly good gas mileage and offer some protection in accidents. This is an example of

(1) a burden of technology on the environment

(2) monitoring short-term and long-term effects of technology

(3) a decision about technology that involves a trade-off between benefits and burdens

(4) modifying technology to reduce or eliminate its drawbacks

29. Although the pesticide DDT was effective in killing insects that damage crops, its use was banned when it was found to be harmful to humans and wildlife. This shows that

    (1) technological products have only disadvantages

    (2) technological products have increased our choices

    (3) use of a technological product may be terminated if its disadvantages outweigh its advantages

    (4) use of a technological product may depend on people's attitudes

Intermediate-Level Science Sample Test 1
Part A: Questions 1–17

Directions (1–17): Each question is followed by four choices. Decide which choice is the best answer. Mark your answer in the spaces provided on the separate answer sheet by writing the number of the answer you have chosen.

1 Which part of a cell allows nutrients and other materials to enter or leave the cell?

  1 cytoplasm
  2 nucleus
  3 chloroplast
  4 cell membrane

2 Which human body system controls production of the hormones that regulate body functions?

  1 digestive
  2 endocrine
  3 respiratory
  4 skeletal

3 Hereditary information is found in a cell's

  1 chloroplasts
  2 chromosomes
  3 cytoplasm
  4 membranes

4 What is a major cause of variation within a species?

  1 sexual reproduction
  2 asexual reproduction
  3 extinction
  4 photosynthesis

5 Which process is shown in the diagram below?

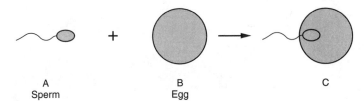

| A | B | C |
| Sperm | Egg | |

  1 metamorphosis
  2 regulation
  3 fertilization
  4 respiration

6 A male chimpanzee has 48 chromosomes in each of his regular body cells. How many chromosomes would be found in each of his sperm cells?

  1 96
  2 48
  3 24
  4 12

7   The diagram below shows materials needed for survival being transported inside a plant.

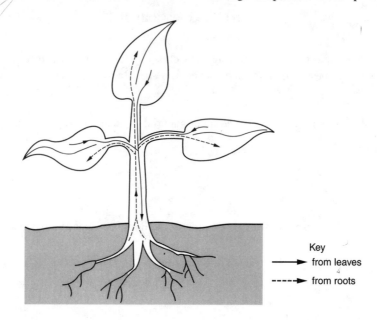

Key
→ from leaves
---→ from roots

Which body system performs this function in humans?

1 circulatory system
2 digestive system
3 excretory system
4 respiratory system

8   The energy content of food is measured in

1 ounces
2 degrees
3 grams
4 Calories

9   The diagram below shows the Moon revolving around Earth as viewed from space.

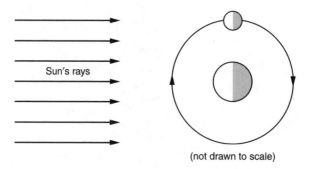

Sun's rays

(not drawn to scale)

What makes it possible to see the Moon from Earth?

1 The surface of the Moon emits its own light, which can be seen from Earth.
2 The Moon absorbs light during the day and emits the light at night.
3 Light emitted by Earth illuminates the Moon's surface, making it visible.
4 Light emitted by the Sun is reflected to Earth by the Moon's surface.

10   The solid part of Earth's surface is called the

   1 hydrosphere
   2 lithosphere
   3 troposphere
   4 atmosphere

11   A rock that contains fossil seashells was most likely formed as a result of

   1 volcanic activity
   2 sedimentation
   3 heat and pressure
   4 magma cooling

12   The diagram below shows the rock cycle.

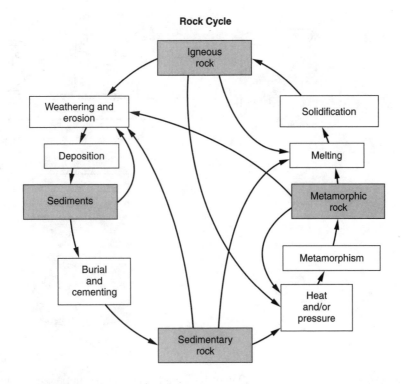

Rock Cycle

Which two processes result in the formation of igneous rocks?

   1 melting and solidification
   2 sedimentation and evaporation
   3 crystallization and cementation
   4 compression and precipitation

13   What process makes a child's breath become visible on a cold day?

   1 boiling
   2 melting
   3 condensation
   4 evaporation

14 In which situation is a chemical reaction occurring?

    1 salt dissolves in water
    2 a nail rusts
    3 ice melts
    4 a glass breaks

15 As ice cream melts, its molecules

    1 absorb heat energy and move farther apart
    2 absorb heat energy and move close together
    3 release heat energy and move farther apart
    4 release heat energy and move closer together

16 Which diagram best shows the property of refraction?

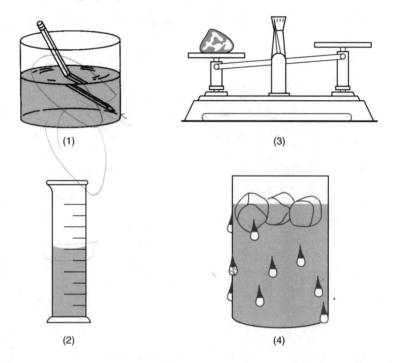

(1)

(3)

(2)

(4)

17 A student pushes against a wall with 20 N of force and the wall does not move. In this situation, the wall exerts

    1 0 N of force
    2 less than 20 N of force
    3 20 N of force
    4 more than 20 N of force

Directions (18–21): Each question is followed by four choices. Decide which choice is the best answer. Mark your answer in the spaces provided on the separate answer sheet by writing the number of the answer you have chosen.

18   The data table below shows the average distance of four planets from the Sun and the approximate time it takes those planets to orbit the Sun.

Planet	Average Distance from the Sun (millions of kilometers)	Approximate Time It Takes the Planet to Orbit the Sun (Earth days)
Mercury	57.9	88
Venus	108.2	225
Earth	149.6	365
Mars	227.9	687

Which statement is best supported by the data in the table?

1 Venus takes less time to orbit the Sun than Mercury does.
2 Mars takes less time to orbit the Sun than Earth does.
3 Mars takes more time to orbit the Sun than Earth does.
4 Venus takes more time to orbit the Sun than Mars does.

19   The data table below shows the masses and volumes of three objects (A, B, and C).

A	B	C
Mass = 4 g Volume = 2 cm³	Mass = 6 g Volume = 6 cm³	Mass = 8 g Volume = 4 cm³

The formula for calculating an object's density is: $\text{Density} = \dfrac{\text{Mass}}{\text{Volume}}$.

Which statement about the densities of these three objects is correct?

1 B is more dense than A.
2 A is more dense than C.
3 B and C have equal densities.
4 A and C have equal densities.

20   The diagram below shows the frequency and wavelength of various types of electromagnetic energy.

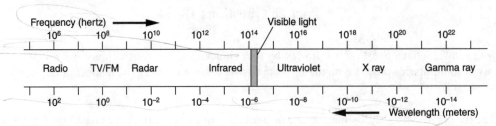

Which type of electromagnetic wave has a wavelength of approximately 10^{-10} meter and a frequency of 10^{18} hertz?

1 infrared
2 radio
3 X ray
4 radar

21   The graph below shows the distance and time traveled by four cars.

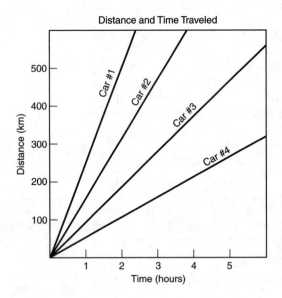

Which car traveled the slowest?

1 Car #1
2 Car #2
3 Car #3
4 Car #4

Directions (22–34): For each question, write your answer in the spaces provided on the separate answer sheet.

Base your answers to questions 22 through 24 on the diagrams and data table below.

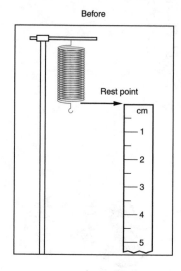

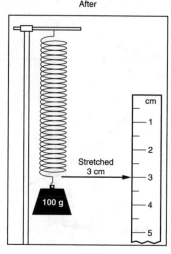

Data table

Attached mass (g)	Distance spring is stretched (cm)
100	3
200	6
300	9
400	12
500	15
1000	30

22   State the relationship between the mass attached to the end of the spring and the length the spring is stretched. [1]

23   Predict how many centimeters the spring will stretch if a total mass of 700 grams were attached. [1]

24   What mass would be needed to stretch the spring to a length of 60 cm? [1]

Base your answers to questions 25 through 27 on the Punnett square and information below.

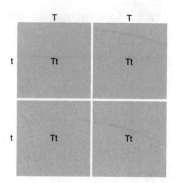

In a certain plant, the gene for tall height (T) is dominant over the gene for short height (t). The Punnett square shows the results of a cross between a pure tall plant and a pure short plant.

25   What percentage of the offspring would be tall plants? [1]

26   Use the Punnett square on your separate answer sheet to show the results of crossing two of the offspring shown in the Punnett square above. [2]

27   Which process is represented by the use of the Punnett square?
1 natural selection
2 sexual reproduction
3 pollination
4 mutation

Base your answers to questions 28 through 32 on the food web shown below.

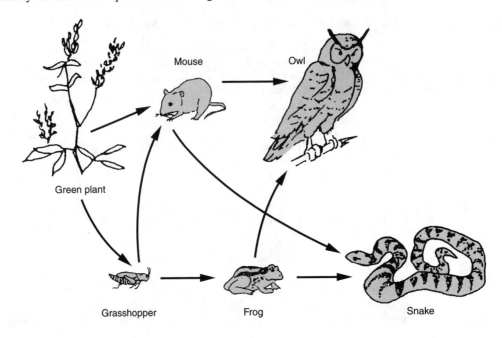

Mouse

Owl

Green plant

Grasshopper

Frog

Snake

28   Identify a producer in this food web. [1]

29   Identify an herbivore in this food web. [1]

30   Identify a carnivore in this food web. [1]

31   Identify an omnivore in this food web. [1]

32   Explain why removing the snake from this food web might result in a decrease in the grasshopper population. [1]

Base your answers to questions 33 and 34 on the diagram below, which shows a form of reproduction.

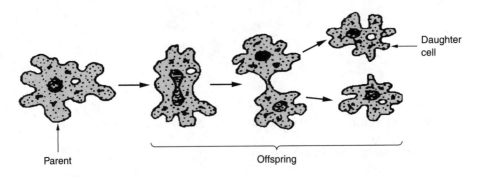

Daughter cell

Parent

Offspring

33   Which type of reproduction is shown in the diagram? [1]

34   How does the genetic material of the daughter cell compare to the genetic material of the parent cell? [1]

Directions (35–45): For each question, write your answer in the space provided on the separate answer sheet.

Base your answers to questions 35 and 36 on the charts below, which show two elements (iron and sulfur) and their properties. The arrows indicate that these elements may combine to form either a mixture of iron and sulfur or the compound iron sulfide.

ELEMENT	PROPERTIES		
Iron	magnetic, black		
Sulfur	nonmagnetic, yellow		

Physical change →

PROPERTIES OF IRON AND SULFUR MIXTURE
Partially magnetic, black and yellow

Chemical change →

PROPERTIES OF IRON SULFIDE COMPOUND
Nonmagnetic, shiny, gray

35   How could a student use a magnet to indicate that combining iron and sulfur to produce the mixture of iron and sulfur is a physical change? [1]

36   What evidence indicates that a chemical change took place when the iron and sulfur combined to form iron sulfide? [1]

Base your answers to questions 37 and 38 on the diagrams below, which show two situations in which energy transformations are occurring.

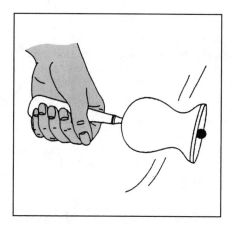

37   As the candle burns, which energy transformation occurs? [1]

38   As the bell rings, which energy transformation occurs? [1]

39   A student plays tennis several times a week. She notices that the tennis ball seems to bounce higher on some courts than on other courts. She wonders if this has something to do with the surface of the court. Design an experiment to see if her hypothesis is correct. Include these elements in your response:

   • State the hypothesis. [1]
   • Identify the factor to be varied. [1]
   • Identify two factors that should be held constant. [2]
   • Clearly describe the procedures. [1]

Base your answers to questions 40 through 42 on the graphs below, which show the laboratory growth of two microorganisms when provided with adequate food and grown in **separate test tubes.**

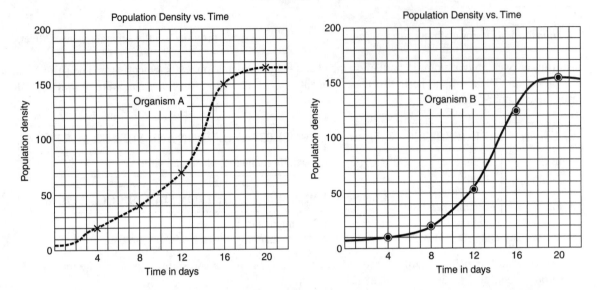

40   The data below were obtained when organism A and organism B were grown with adequate food in the **same test tube**.

Population density	Day					
	0	**4**	**8**	**12**	**16**	**20**
Organism A	5	50	100	75	25	10
Organism B	5	25	75	125	150	125

On the grid provided on your separate answer sheet, make a graph of the data from the table above according to the instructions below. [3]

a Place an *X* to show the population of organism A for each interval in the 20-day period.
b Connect the *X's* with a *dashed* line. Make a key that indicates this line represents the data for organism A.
c Place a dot in a circle to show the population of organism B for each interval in the 20-day period.
d Connect circled dots with a *solid* line. Make a key that indicates this line represents the data for organism B.

41   State a relationship that may have produced the results shown when organism A and organism B were grown together. Explain your answer, using the graphed data. [2]

42   Based on your graph, predict the population density of organism A *or* organism B at day 21. Explain your prediction. [2]

Base your answers to questions 43 through 45 on the information below and on your knowledge of science. The page of notes shown below was made by a student doing a research project about hail.

Some Observations about Hail:

1 Hailstones are fairly round in shape.
2 When cut in half or held up to the light, layers can be seen in the hailstone.
3 Average diameter of hailstones found after last summer's storms:
  July 12—6 mm
  July 26—9 mm
  August 12—20 mm
  August 26—12 mm

Background Information Found in Science Book:

1 The precipitation dropped during a hailstorm is called hailstones.
2 Hail causes damage to crops, buildings, and vehicles.
3 Hailstones are usually more than 5 mm in diameter.
4 Thunderstorms form from tall clouds which have temperatures below 0°C in their upper regions. They have strong updrafts pushing the air toward the top.
5 The layers in a hailstone are caused by the path the hail takes as it falls.
6 A new layer forms each time the hailstone is pushed into the freezing zone.

43 On the diagram on your separate answer sheet, draw a path that would produce a hailstone that has three layers. You can practice on the drawing below. Be sure your path starts and finishes at the points shown. The dotted line separates the thunderstorms in Zone A, which is above 0°C, and Zone B, which is below 0°C. When you are satisfied with your path, copy it onto the diagram on your answer sheet. Your answer will be evaluated by how well the path you have drawn could produce a three-layer hailstone. [2]

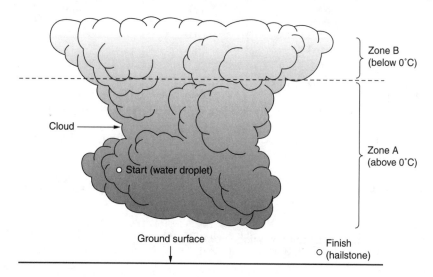

44 Based on the student's observations above, which of last summer's storms most likely had the strongest updrafts associated with it? Give two reasons to support your answer. [3]

45 Based on the background information in the student's notes above, identify one problem that can be caused by hailstorms. [1]

Part D—Performance Test
Station X: Experimenting with a Ball and Ramp

> **TASK:**
> At this station, you will observe a ball rolling down a ramp and moving a plastic cup. You will then identify some variables that would affect how far the cup moves. Finally, you will design an experiment and formulate a hypothesis.

DO NOT MOVE THE RAMP SETUP

DIRECTIONS:

1   Be sure the cup is on the Starting Circle, with the opening in the cup facing the end of the ruler, at the start of each trial.

2   Take the golf ball out of the bag.

3   Place the ball on the ramp (ruler) so the middle of the ball is at the 15.0-cm mark on the ruler. Without pushing the ball, carefully release it so the ball goes into the cup. Note the distance the cup moves.

4   On data table 1 below, record the distance the cup moves when the ball is released from the 15.0-cm mark. Record the distance to the nearest 0.1 cm.

DATA TABLE 1

Ball's Release Point (cm)	Mass of Cup (g)	Distance Cup Moved (cm)
15.0	16.0	
20.0	16.0	
25.0	16.0	
25.0	16.0	
25.0	16.0	

5   With the cup returned to the Starting Circle, release the golf ball from the 20.0-cm mark on the ruler. Record the distance the cup moves to the nearest 0.1 cm in data table 1.

6   With the cup returned to the Starting Circle, release the golf ball from the 25.0-cm mark on the ruler. Record the distance the cup moves to the nearest 0.1 cm in data table 1.

7   With the cup returned to the Starting Circle each time, release the golf ball two more times from the 25.0-cm mark on the ruler. Each time, record the distance the cup moves to the nearest 0.1 cm in the data table 1.

8   You probably found that the cup traveled slightly different distances when you released the ball three times from the 25.0-cm mark. Give two reasons that might explain why the cup did *not* stop at the *exact* same spot each time.

First Reason: _____

_____

Second Reason: _____

_____

9　Think about how you might design a new experiment. In this experiment you want to study how changing the mass of the cup will change the distance it is moved by the golf ball. Assume that the equipment setup for this new experiment will be the same as what is now at your station. The data table for this new experiment is shown below. (**Do *not* actually fill in data table 2.**)

DATA TABLE 2

Ball's Release Point (cm)	Mass of Cup (g)	Distance Cup Moved (cm)
	20.0	
	40.0	
	60.0	
	80.0	

10　What would you recommend about the release point of the golf ball each time a new cup is tested?

_____

_____

_____

11　Write a hypothesis about the distance the cups of different masses will be moved by the golf ball.

_____

_____

_____

_____

12　Return all materials to their positions as shown on the Station Diagram.

> **STOP**

Station Y: Soaps and Water

> **TASK:**
> At this station, you will determine some properties of two soap samples and predict how they would behave if they were placed in water. You will then place two objects in water and compare their densities.

DIRECTIONS:

1 To protect the soap samples, do NOT take them out of the plastic bags and do NOT place the soaps in water. Disregard the effect of the plastic bags for all measurements and calculations.

2 What is the number on the bag for Soap A? _____

   What is the number on the bag for Soap B? _____

3 Use the data table below to record your answers to questions 4–7.

SOME PROPERTIES OF TWO SOAP SAMPLES

Soap	Mass (g)	Volume (cm³)	Density (g/cm³)
A B			0.8

4 Measure the mass of Soap A and measure the mass of Soap B. Record the values to the nearest 0.1 g for each in the data table above. (Note that the unit, g (grams), has been provided.)

5 Measure the length, width, and height of Soap A to the nearest 0.1 cm. Record these dimensions in the work space below. Substitute your values in the formula provided. Then use the calculator to determine the volume of Soap A. Show your work in the space below. Record your value to the nearest 0.1 cm³ in the data table.

> Work Space
>
> Length _____ cm   Width _____ cm   Height _____ cm
>
> Volume = Length × Width × Height

6 For Soap A, substitute your values for mass and volume in the formula provided. Then use the calculator to determine the density of Soap A. Show your work in the space below. Record your value to the nearest 0.1 g/cm³ in the data table.

> Work Space
>
> Density = $\dfrac{\text{Mass}}{\text{Volume}}$

7 Notice that the density of Soap B has been provided in the data table (0.8 g/cm³). For Soap B, substitute your values for mass and density in the formula provided. Then use the calculator to determine the volume of Soap B. Show your work in the space below. Record your value to the nearest 0.1 cm³ in the data table on page 346.

Work Space

$$\text{Density} = \frac{\text{Mass}}{\text{Volume}}$$

8 The diagram below represents a glass container with water. Think about what would happen if Soap A and Soap B were removed from the plastic bags and placed in this container. **Remember, do *not* actually put the soaps in the water**.

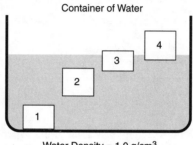

Container of Water

Water Density = 1.0 g/cm³

Base your answers to questions 9 and 10 on the values in your data table on page 346. (Circle one answer for each question.)

9 Which block in the diagram above shows about where Soap A would be if it were placed in the container of water?

(Circle one)   Block 1   Block 2   Block 3   Block 4

10 Which block in the diagram above shows about where Soap B would be if it were placed in the container of water?

(Circle one)   Block 1   Block 2   Block 3   Block 4

___

11 Take the rubber ball and the Styrofoam ball out of the bag. Place them in the plastic cup with water. Observe the position of the balls in the water. Based on your observations, how does the density of the rubber ball compare with the density of the Styrofoam ball? Explain your answer.

12 Remove the balls from the cup, wipe them off, and return them to the bag.

13 Return all materials to their positions as shown on the Station Diagram.

STOP

Station Z: Cell Size

TASK:
At this station you will measure the size of a microscope's field of view, estimate the size of a cell, and draw pictures of cells that you observe under the lowest and highest powers.

DIRECTIONS:

1   Pick up Slide A, hold it up to the light, and look at the squares.

2   Slide A is a prepared slide of a tiny piece of graph paper.

    The lines of the graph paper are all spaced 1.0 mm apart.

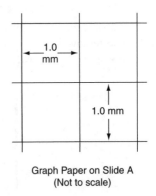

Graph Paper on Slide A
(Not to scale)

3   Place Slide A on the microscope stage and bring the graph paper into focus, using the lowest power.

4   When you look into the microscope, the whole area you see is called "field of view." Knowing that the lines of the graph paper are 1.0 mm apart, estimate the diameter of the lowest power's field of view to the nearest 0.25 mm.

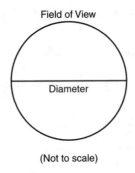

Field of View

Diameter

(Not to scale)

Estimated diameter of the lowest power's field of view: _____ mm

5   Return Slide A.

6   Place Slide B on the microscope and bring it into focus under the lowest power. Slide B is a piece of onion skin tissue that has been stained and mounted for viewing. See the diagram below for a sketch of what one cell might look like. The cell length has been labeled.

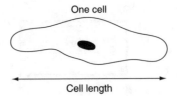

One cell

Cell length

7   Look closely at Slide B under the lowest power. Find one row of cells that goes across the middle of the field of view from one edge of the field of view to the other edge. These cells may go from side to side, from top to bottom, or diagonally across the diameter. In the circle at the right, carefully sketch only *one row* of cells whose lengths go across the field of view.

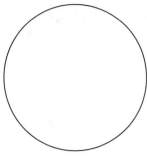

One Row of Cells Under Lowest Power

8   How many cells did you see under lowest power in the row that you drew above? ____ cells

9   In Step 4 on the previous page, you estimated the diameter of the lowest power's field of view. Record that value again here: _____ mm

10 Based on the values you recorded in Steps 8 and 9, calculate the average length of *one* onion cell in your diagram to the nearest 0.1 mm. _____ mm/cell

    Work Space

11 Return Slide B.

12 Place Slide C on the microscope stage. Bring Slide C into focus under the lowest power. Now bring the slide into focus under the highest power. In the box below, draw an enlarged view of one typical cell on this slide under the highest power. Your drawing should accurately show the shape and structures of the cell.

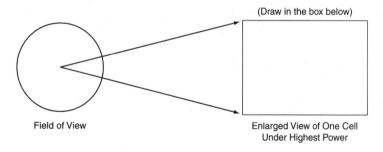

(Draw in the box below)

Field of View

Enlarged View of One Cell
Under Highest Power

13 When you are finished, put the microscope back to the lowest power. Return all materials to their positions as shown on the Station Diagram.

STOP

Intermediate-Level Science Sample Test 1
Student Answer Sheet for Parts A, B, and C

Student Name _____

Directions (1–21): Each question is followed by four choices. Decide which choice is the correct answer. Mark your answer in the spaces below by writing the number of the answer you have chosen. (NOTE: A scannable answer sheet will be provided for the actual exam. This format is provided for the sample test only.)

1 _____	8 _____	15 _____
2 _____	9 _____	16 _____
3 _____	10 _____	17 _____
4 _____	11 _____	18 _____
5 _____	12 _____	19 _____
6 _____	13 _____	20 _____
7 _____	14 _____	21 _____

Directions (22–39): For each question, write your answer in the spaces below.

22 _____

23 _____ cm

24 _____ grams

25 _____ %

26

27 _____

28 _____

29 _____

30 _____

31 _____

32 _____

_____

33 _____

34 _____

_____

35 _____

_____

36 _____

37 _____

38 _____

39 Hypothesis: _____

_____

Two factors to be held constant: _____

    1 _____

    2 _____

Procedure: _____

_____

_____

_____

_____

_____

_____

40

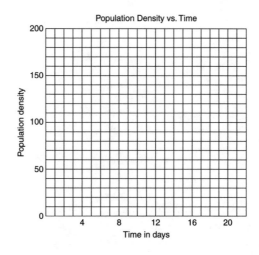

41 _____

_____

42 _____

_____

43

44 _____

_____

_____

45 _____

Intermediate-Level Science Sample Test 2
Part 1: Questions 1–35

Directions (1–35): Each question is followed by four choices. Decide which choice is the best answer. Mark your answer in the spaces provided on the separate answer sheet by writing the number of the answer you have chosen.

1 Cells use oxygen to release energy from food in the process known as

   1 excretion
   2 locomotion
   3 digestion
   4 respiration

2 Which part of the cell controls cell activities and stores genetic material?

   1 nucleus
   2 cell membrane
   3 chloroplast
   4 centriole

3 Which life process is necessary for the survival of a species but not necessary for the survival of the individual organism?

   1 excretion
   2 nutrition
   3 respiration
   4 reproduction

4 Which of the following is a main function of the nervous system?

   1 transport materials throughout the body
   2 break food down into a usable form
   3 control and coordinate response to stimuli
   4 producing sex cells

5 Which organism is best viewed under a microscope?

   1 paramecium
   2 elephant
   3 earthworm
   4 grasshopper

6 Which of the following structures are found only in plant cells?

   1 chromosomes
   2 chloroplasts
   3 cell membranes
   4 cytoplasm

7 Green plants are called producers because

   1 they produce oxygen gas
   2 they obtain energy from animals
   3 they manufacture their own food
   4 they can be decomposed by bacteria

8 Asexual reproduction differs from sexual reproduction in that in asexual reproduction

   1 the offspring are genetically different from either of their parents
   2 the offspring are genetically identical to both of their parents
   3 the offspring are genetically different from their single parent
   4 the offspring are genetically identical to their single parent

9 According to the system of biological classification, human beings, cats, and elephants are placed in the same

   1 species
   2 genus
   3 family
   4 class

10 Which process is illustrated in the diagram below?

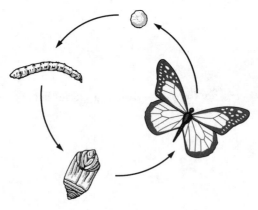

Caterpillar — Cocoon ⟶ Butterfly

   1 regulation
   2 fertilization
   3 metamorphosis
   4 reproduction

11 A male fruit fly has 4 chromosomes in each of its sperm cells. How many chromosomes would be found in each of its regular body cells?

   1 8
   2 2
   3 16
   4 4

12 Tall pea plants that are crossed with short pea plants produce offspring that are all tall. Which would be the best conclusion to make about the inheritance of height in pea plants.

   1 height cannot be inherited
   2 tall is dominant
   3 tall is recessive
   4 short is dominant

13 Infectious diseases are caused by

   1 damaged organs
   2 poor diet
   3 heredity
   4 microorganisms

14 Hay fever is a disease in which pollen in the air can cause itchy eyes and a stuffy nose. This is an example of a disease caused by

1 environmental factors
2 bacteria
3 viruses
4 poor diet

15 Herbivores obtain energy directly from

1 the sun
2 plants
3 animals
4 decaying organisms and wastes

16 Which process is illustrated in the cartoon?

1 melting, with a release of energy
2 freezing, with a release of energy
3 melting, with an absorption of energy
4 freezing, with an absorption of energy

17 Which of the following is a physical change?

1 burning of wood
2 boiling of water
3 rusting of iron
4 cooking an egg

18 When water is heated from 20°C to 40°C, its molecules

1 move faster and get farther apart
2 move faster and get closer together
3 move slower and get farther apart
4 move slower and get closer together

19 Which of the following is classified as a compound?

1 hydrogen
2 water
3 iron
4 salt water

20 Which of the following resources is considered nonrenewable?

    1 solar energy
    2 wind energy
    3 moving water
    4 fossil fuels

21 What makes it possible for an astronaut to see Earth from space?

    1 Earth emits light from its surface.
    2 The moon emits light, and this light is reflected from Earth.
    3 The sun emits light, and this light is reflected from Earth.
    4 Sun emits light, and this light is reflected from the moon.

22 Ice water placed in a dry glass on a warm summer day will cause water to appear on the outside of the glass. Where did the water on the outside of the glass come from?

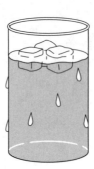

    1 the ice
    2 the water in the glass
    3 the water vapor in the air
    4 a crack in the glass

23 What type of weather conditions would be expected in New York State from an air mass originating in the of Gulf of Mexico?

    1 cool and dry
    2 cool and moist
    3 warm and dry
    4 warm and moist

24 Earthquakes, volcanic eruptions, and the creation of mountains and ocean basins are events that occur in response to

1 glacial activity on Earth's surface
2 heat flow from within Earth's interior
3 ocean currents
4 pressure differences between the lithosphere and atmosphere

25 The solid portion of Earth's surface is called the

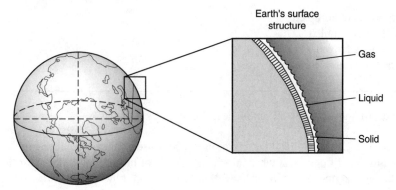

1 hydrosphere
2 lithosphere
3 atmosphere
4 troposphere

26 Driving a car is an example of a feedback system. The following diagram shows how the components in a car system work together. The feedback in this system is provided by the

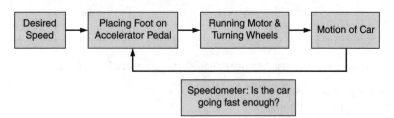

1 accelerator
2 motor
3 wheels
4 speedometer

27 A closed bottle of water was left outdoors on a very cold night when the temperature was −5°C. The next morning the water was frozen, and the bottle was cracked. Why was the bottle cracked?

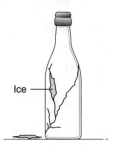

1 Water contracts when it freezes.
2 Water expands when it freezes.
3 All substances contract when they freeze.
4 Water releases heat energy when it freezes.

28 Newton's second law of motion states the relationship among force, mass, and acceleration. The law is expressed in the following formula:

$$\text{acceleration} = \frac{\text{Force}}{\text{mass}}, \text{ or } a = \frac{F}{m}$$

For an object in motion, what happens to the acceleration of the object if the force is doubled and the mass remains constant?

1 acceleration is quadrupled
2 acceleration is doubled
3 acceleration is halved
4 acceleration remains the same

Base your answers to questions 29–32 on the description of the lab.

One day in science lab, Omar decided to test five substances to see if they were living organisms. He put a small sample of each into five different test tubes, along with 3 mL of bromthymol blue. Test tube 6 is his control; it contains only the bromthymol blue. Omar knew that bromthymol blue turns yellow in the presence of carbon dioxide gas. He corked all the tubes and waited 24 hours. Here is a table of his results.

Test Tube	Substance	Color of Bromthymol Blue
1	Yogurt	Dark yellow
2	Salt	Dark blue
3	Yeast	Yellow-orange
4	Bean seeds	Dark yellow
5	Sand	Dark blue
6	—	Dark blue

29 According to this table, which substances should Omar conclude are living?

1 yogurt, yeast, and bean seeds
2 salt, sand, and bromthymol blue
3 yogurt, salt, and sand
4 yeast, bean seeds, and sand

30 To come to this conclusion, what inference must Omar make about living things?

1 They give off carbon dioxide from respiration.
2 They have the element carbon in their cells.
3 They are able to reproduce in bromthymol blue.
4 They absorb the blue nutrients from the liquid, leaving the yellow.

31 How do these living cells get energy for growth?

1 They all absorb it from the sun.
2 They all absorb it from the bromthymol blue.
3 They all chemically change food into energy.
4 They don't have to; it doesn't take any energy to grow.

32 While cleaning up at the sink, Omar broke one of his test tubes. Which action should he have taken?

1 quickly pick up the larger pieces with his hands
2 use a broom to sweep all the pieces into the trash can
3 break it again until all the pieces are small enough to be rinsed down the sink
4 not touch any piece but tell the teacher about it

33 A mixture of salt, sand, and water is thoroughly stirred. It is then filtered through a filter paper as shown in the diagram below. Which substance(s) pass through the filter?

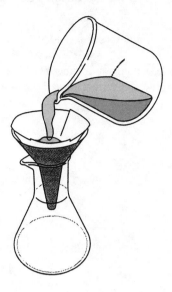

1 only the water
2 only the water and the sand
3 only the water and the salt
4 the sand, the salt and the water.

Base your answers to questions 34–35 on the diagram below, which represents a beetle and a maple tree.

34 What can the beetle do that the tree cannot do?

1 make its own food
2 move from place to place
3 reproduce
4 take in oxygen to burn food for energy

35 What can the maple tree do that the beetle cannot do?

1 make its own food
2 move from place to place
3 reproduce
4 take in oxygen to burn food for energy

Directions (36–72): Write your answer in the spaces provided on the separate answer sheet.

36   Which diagram best represents how a person in New York State would observe the rising of a group of three stars? Why?

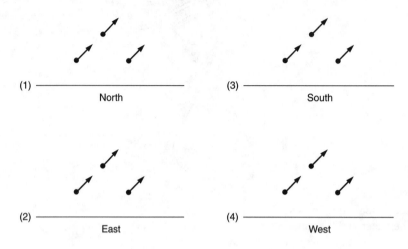

(1) —————————————   (3) —————————————
            North   South

(2) —————————————   (4) —————————————
             East   West

Base your answers to questions 37 through 39 on the information below.

The mass and volume of an aluminum block were measured and recorded in the Table A below. Table B gives the density of 3 liquids.

TABLE A

Mass = 27.0 g
Volume = 10.0 cm³

Table B

Substance	Density
Water	1.0 g/cm³
Ethanol	0.8 g/cm³
Mercury	13.6 g/cm³

37   What is the density of this aluminum block? (Show all work on the answer sheet.)

38   In which of the liquids in Table B would this aluminum block float?

39   How would the density of a 50.0 cm³ aluminum block compare with that of the block described in Table A above.

Base your answers to questions 40 and 41 on the chart below.

PROPERTIES OF THE SUBATOMIC PARTICLES

Particle	Mass (AMU)*	Charge	Location
Proton	1	+	Nucleus
Neutron	1	0	Nucleus
Electron	0.00054	−	Outside the Nucleus

*Atomic Mass Unit

40 A certain particle contains 3 protons, 4 neutrons, and 2 electrons. To the nearest whole number, what is the mass of this particle?

41 What is the charge of the particle in question 40?

Questions 42–44 are based on the information below that represents a food pyramid.

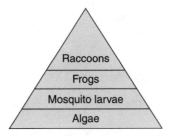

42 State one reason that algae form the base of this pyramid.

43 What position does the mosquito larvae occupy in this food pyramid?

44 A disease kills most of the frogs in this ecosystem. What would happen to the populations of raccoons and mosquitoes?

---

45 What structures do the cells of a maple tree and the cells of a beetle have in common?

46 Fruit flies can have red or white eyes. When studying the inheritance of this trait, biologists use the letter "R" to represent the gene for red eyes, and "r" to represent the gene for white eyes. What would a fly with the gene combination "Rr" be?

Base your answers to question 47 on the following information: According to Newton's First Law of Motion, an object in motion will remain in motion until an outside force acts on the object.

47   State two forces that cause a ball that was thrown upward to fall to the ground and stop.

48   Design a demonstration of Newton's Third Law of Motion—For every action there is an equal and opposite reaction.

Base your answers to questions 49–51 on the diagram below, which represents the four inner planets at various positions in their orbits.

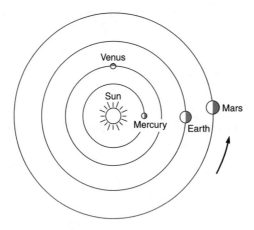

49   Why is Mars visible from Earth in the night sky?

50   Why is Mercury not visible from Earth?

51   At what time of day, is Venus visible from Earth?

---

52   In air at 20°C, sound travels at 343 m/s. As the temperature of air rises, the speed of sound increases at the rate of 0.6 m/s per degree C. What is the speed of sound at 30°C? (Show all work on your answer sheet.)

Base your answers to questions 53 and 54 on the following information. The diagram shows white light hitting a surface. The red, orange, yellow, green and violet portion of the light is absorbed by the surface.

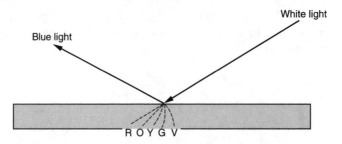

53   What happened to the blue portion of the light?

54   What color will the surface appear to be?

Base your answers to questions 55–60 on the data tables below, which represent the average heights of a group of pea plants watered with Fertilizer A and a group of pea plants watered with Fertilizer B. Using the information in the data tables, construct a line graph on the grid provided on the answer sheet. Follow the direction below.

FERTILIZER A

Day	Height
0	2
2	10
4	15
6	18
8	20
10	21
12	21

FERTILIZER B

Day	Height
0	2
2	5
4	8
6	11
8	14
10	17
12	20

55 Plot each point for Fertilizer A and surround each point with a small circle. Connect the points with a solid line.

Example of points to plot:

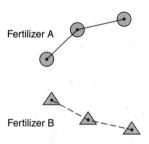

Fertilizer A

Fertilizer B

56 Plot each point for Fertilizer B and surround each point with a small triangle. Connect the points with a dashed line.

57 What would be an appropriate title for this graph?

58 Based on the graph, what would you predict about the heights of the plants with Fertilizer A compared to the heights of the plants with Fertilizer B on day 14?

For questions 59 and 60, complete the table on your answer sheet as follows.

List 2 environmental factors that should be kept the same for each plant being tested in this experiment. (2 points)

For each factor you list, give a scientific reason why the factor needs to be the same for all of the plants. (2 points) You may use the table below for practice.

	Factor to be kept constant	Reason
59		
60		

Base your answer to question 61 on the map below.

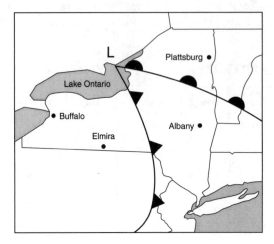

61   Describe the weather change for the past six hours in Elmira, New York.

62   When the switch is closed in the electrical circuit illustrated below, the lightbulb did not light. Explain why the bulb did not light.

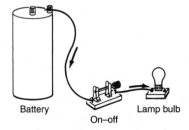

Base your answers to questions 63 and 64 on the diagram.

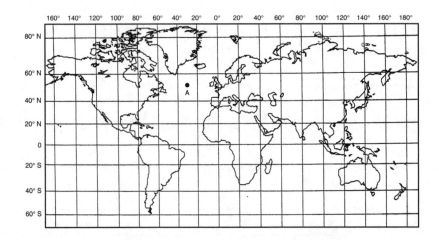

63   What is the latitude and longitude of location A, a volcano on the ocean floor?

64   With what geologic feature is this volcano associated?

Base your answers to questions 65–67 on the diagrams below, which show three sealed containers, labeled A, B, and C.

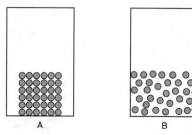 

The circles in container A represent the molecules of a substance in the solid state.
The circles in container B represent the molecules of the same substance after it has been melted.

65   In container C, draw a diagram showing the molecules of this substance in the gaseous state.

66   What must be done to the molecules in container B to convert them to the gaseous state?

67   Changing a substance from one state to another, is considered a physical change. Explain why this is a physical change and not a chemical change.

Base your answers to questions 68–70 on the geologic cross section below. The order of events that created this geologic cross section are (1) deposition of sedimentary rocks, (2) igneous rock intrusion, and (3) faulting.

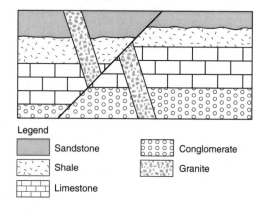

Legend

Sandstone   Conglomerate
Shale   Granite
Limestone

68   (a) Which rock layer formed first?
     (b) How did you determine this?

69   What evidence indicates that the granite formed after the limestone?

70   In the diagram on your answer sheet, draw two arrows to show the relative motion of the rocks along the fault line. You may use the diagram here for practice.

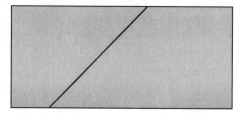

Base your answers to questions 71 and 72 on the world map below.

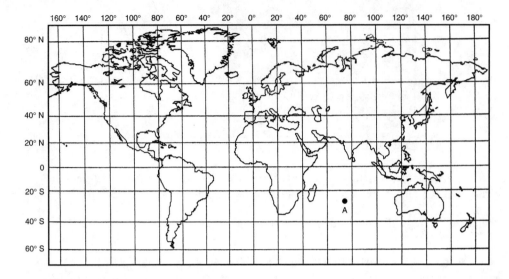

71   What is the latitude and longitude of Point A?

72   Place an "X" on the map on your answer sheet to indicate the location of an earthquake at latitude 40°
      N and longitude 120° W.

Intermediate-Level Science Sample Test 2
Student Answer Sheet for Parts 1 and 2

Student's Name _____

Directions (1–35): Each question is followed by four choices. Decide which choice is the correct answer. Mark your answer in the spaces below by writing the number of the answer you have chosen. (NOTE: A scannable answer sheet will be provided for the actual exam. This format is for the sample test only.)

1 _____	13 _____	25 _____
2 _____	14 _____	26 _____
3 _____	15 _____	27 _____
4 _____	16 _____	28 _____
5 _____	17 _____	29 _____
6 _____	18 _____	30 _____
7 _____	19 _____	31 _____
8 _____	20 _____	32 _____
9 _____	21 _____	33 _____
10 _____	22 _____	34 _____
11 _____	23 _____	35 _____
12 _____	24 _____	

Intermediate-Level Science Sample Test 2
Student Answer Sheet for Part 2

Directions (36–72): For each question, write your answer in the spaces below.

36 _____

37

_____

38 _____

39 _____

40 _____

41 _____

42 _____

_____

43 _____

_____

44 _____

_____

45 _____

46 _____

_____

47 _____

_____

48 _____

_____

_____

49 _____

_____

50 _____

_____

51 _____

_____

52

_____

53 _____

54 _____

55 and 56

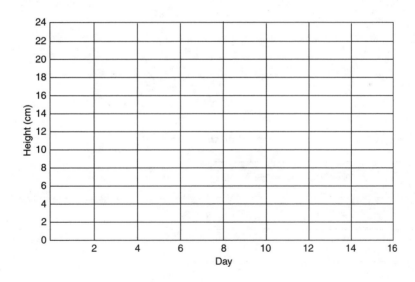

57 _____

58 _____

	Factor to be kept constant	Reason
59		
60		

61 _____

62 _____

63 _____

64 _____

65

c

66 _____

67 _____

68 (a) _____

_____

(b) _____

_____

69 _____

70

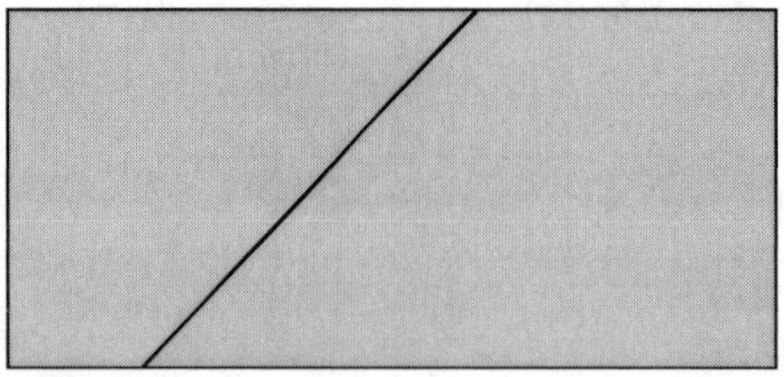

71 _____

72

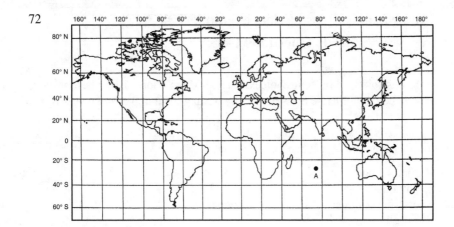

Directions (1–35): Each question is followed by four choices. Decide which choice is the best answer. Mark your answer in the spaces provided on the separate answer sheet by writing the number of the answer you have chosen.

1   Which of the following is an example of mechanical digestion?

1 Wastes are removed from the body.
2 Food is chewed in the mouth.
3 Proteins are broken down into simpler substances by enzymes.
4 Food is "burned" in the cell to produce energy.

2   Which of the following processes is correctly paired with its function?

1 Circulation—releases the energy stored in food
2 Digestion—produces sex cells necessary for the production of offspring
3 Excretion—disposes of liquid and gaseous wastes
4 Respiration—moves substances to and from the cells

3   Of the following groups, which is organized from simple to complex?

1 tissue → cell → organ system → organ
2 organ system → organ → cell → tissue
3 organ → tissue → organ system → cell
4 cell → tissue → organ → organ system

4   Japanese Beetles and June Bugs are classified in the same order. Therefore, they must also be in the same

1 kingdom
2 family
3 genus
4 species

5   Biodiversity refers to the variety of living species. It is important to have a great biodiversity because it provides humans with

1 diversity of food sources
2 diversity in sources of medicines
3 industrial diversity
4 all of the above

6   The union of an egg cell and a sperm cell is called

1 mitosis
2 meiosis
3 asexual reproduction
4 fertilization

7   Which specialized cells protect the body from infectious disease?

1 red blood cells
2 white blood cells
3 nerve cells
4 muscle cells

8 All giraffes have long necks, which enable them to reach leaves in tall trees. Why might there be no short-necked giraffes?

1 Short-necked giraffes could not see as far and were easier prey for predators.
2 Long-necked giraffes eat better, so they have more energy to grow long necks.
3 Short-necked giraffes could not compete with long-necked giraffes to reach leaves in tall trees.
4 In the past, female giraffes only mated with long-necked male giraffes.

9 All of the organisms that live in a pond make up the

1 habitat
2 community
3 environment
4 ecosystem

10 In which environment would you most likely find an animal with thick fur?

1 desert
2 tropical rain forest
3 Arctic tundra
4 grassland

11 Which of the following particles is positively charged?

1 proton
2 electron
3 neutron
4 molecule

12 Which of the following substances would dissolve fastest?

1 powdered sugar in cold water
2 sugar cubes in cold water
3 powdered sugar in hot water
4 sugar cubes in hot water

13 The tool shown below can be used to

1 observe planets that are far away
2 determine the mass of an object
3 view a single cell
4 measure the volume of a liquid

14   The diagrams represent magnified crystals of an igneous rock. Which igneous rock most likely was produced by magma that cooled the quickest and closest to Earth's surface?

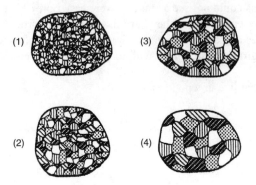

15   Which of the following substances is an element?

1 water
2 air
3 iron
4 table salt

16   Which is an example of an infectious disease?

1 sickle cell anemia
2 lung cancer
3 iron-poor blood
4 AIDS

17   The presidential election of 2000 demonstrated problems in our voting process. The problem demonstrates a relationship between

1 science and society
2 science and technology
3 society and technology
4 long-term and short-term effects of technology

18   Which block diagram represents the mid-Atlantic Ridge?

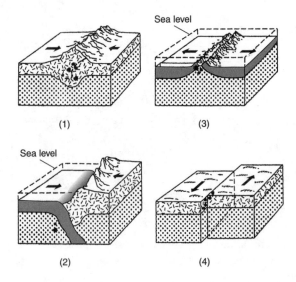

19   What type of energy is demonstrated by a sled speeding downhill?

1 kinetic
2 potential
3 chemical
4 light

20   How is heat transferred when ice is placed in a glass of soda?

1 from the soda to the ice
2 from the ice to the soda
3 from the ice to the glass
4 from the ice to the air

21   On January 3, Earth is located at position C in the diagram. At what point in its orbit will Earth be six months later?

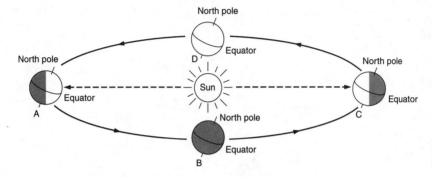

1 A
2 B
3 C
4 D

22   As you travel from an altitude of 0 kilometers to an altitude of 80 kilometers, the temperature

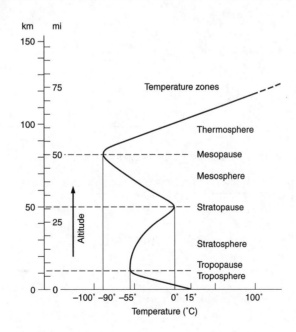

1 increases
2 decreases
3 remains the same
4 decreases, increases, and decreases

23   A screw is a modified form of

1   a wheel and axle
2   a lever
3   a pulley
4   an inclined plane

24   Which of the following processes can separate dissolved salt from water?

1   filtration
2   stirring
3   evaporation
4   sedimentation

25   Which is a chemical change?

1   melting
2   evaporation
3   condensation
4   burning

26   If Earth were tilted 35° from the vertical rather than 23.5° the following change would occur.

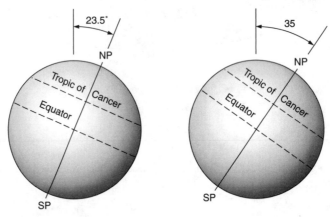

1   the equator would be farther north
2   the Tropic of Cancer would be farther north
3   the equator would be farther south
4   the Tropic of Cancer would be farther south

Use the following weather station information to answer question 27.

Temperature	20°C
Air pressure	29.82 inches of mercury
Wind	Southeast 5–10 miles per hour
Relative humidity	100%
Cloud cover	100%

27   What type of precipitation is most likely occurring?

1   rain
2   snow
3   hail
4   none

28 Which of the following chemical principles is illustrated by the cartoon below?

1 Like charges attract.
2 Opposite charges attract.
3 Like charges repel.
4 Opposite charges repel.

Base your answers to questions 29–30 on the diagram below.

29 The screwdriver being used to remove the lid from a can of paint is acting as

1 a wedge
2 a lever
3 an inclined plane
4 a screw

30 Compared with the amount of work put into a machine, the amount of work put out by the machine is

1 always greater
2 always less
3 always the same
4 sometimes greater, sometimes less

Base your answers to questions 31–32 on the information below.

There are four human blood types: A, B, AB, and O. Blood types are determined by observing the reactions of blood with two antibodies called anti-A and anti-B. The reaction with another antibody called anti-D tells if the blood type is positive or negative for another substance. A person with type A blood produces anti-B antibodies, and a person with type B blood produces anti-A antibodies. A person with AB produces neither anti-A nor anti-B antibodies. A person with type O blood produces both anti-A and anti-B antibodies. A person with negative blood produces anti-D antibodies. The charts below show the blood types.

| Type | Reacts With: | |
	Anti-A	Anti-B
A	yes	no
B	no	yes
AB	yes	yes
O	no	no

	Reaction With Anti-D
Yes	Positive
No	Negative

31 A patient's blood reacts with anti-A and anti-D but not with anti-B. His blood type is

1 A positive
2 A negative
3 B positive
4 O positive

32 A patient at the hospital needs blood. Her body produces anti-A, anti-B, and anti-D. To keep her body from reacting with the new blood, which type of blood should the doctors give her?

1 A positive
2 B negative
3 AB positive
4 O negative

Use the diagram below, which represents a rock cross section to answer questions 33-35.

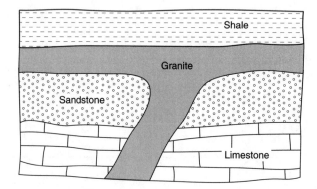

33 Which age relationship is correct?

1 The shale is older than the sandstone.
2 The granite is older than the sandstone
3 The limestone is older than the sandstone
4 The granite is older than the limestone.

34 The granite most likely formed by

1 sedimentation
2 volcanic intrusion underground
3 volcanic extrusion on Earth's surface
4 cementation

35 The limestone most likely formed by

1 sedimentation
2 volcanic intrusion underground
3 volcanic extrusion on Earth's surface
4 cementation

Directions (36–72): For each question, write your answers in the spaces provided on the separate answer sheet.

Base your answers to questions 36 and 37 on the weather map

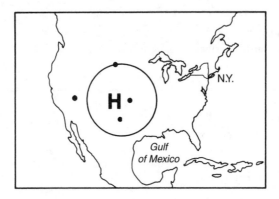

36   What does the **H** on the weather map indicate?

37   As the air mass moves, what section of the United States will it most likely affect in the next few days?

38   Describe what happens when the north pole of magnet A is brought close to the north pole of the suspended magnet B

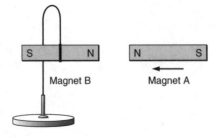

Read the following paragraph and answer questions 39 and 40.

Gail and Jared set up an experiment that they saw in their science book. They used a drinking glass, several different coins (penny, nickel, and quarter), and an index card as shown in the diagram. They flicked the card from under each of the coins to see what would happen.

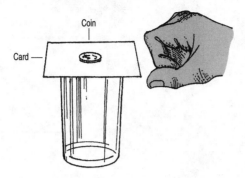

39   Describe what happened as they flicked the card from under each of the coins.

40   The results of this experiment indicated that an object at rest will remain at rest, and will resist being put into motion. What property of matter did this experiment demonstrate?

41   In an organism, what is the main function of blood?

Use the information below to answer questions 42–45

The Punnett square below represents a cross between a yellow and a green pea plant.

	y	y
Y	Yy	
Y		

Key
Y = gene for yellow
y = gene for green

42   Using the Punnet square on your answer sheet, complete the results of the cross between these two pea plants.

43   What percentage of the offspring would be green?

44   Classify the genetic makeup of the parent plants.

45   Define the term hybrid.

Base your answers to questions 46-49 on the following diagram, which shows a food web.

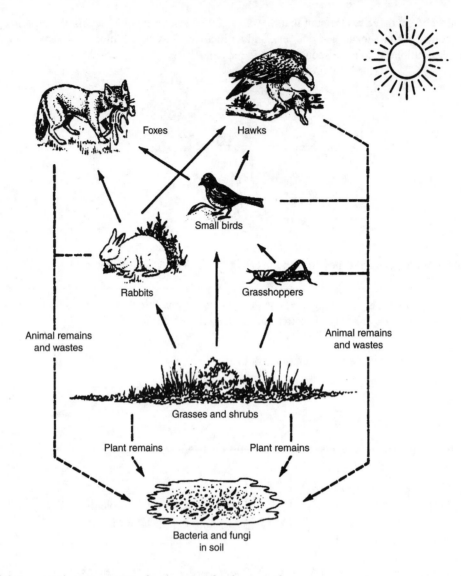

Foxes

Hawks

Small birds

Rabbits

Grasshoppers

Animal remains and wastes

Animal remains and wastes

Grasses and shrubs

Plant remains

Plant remains

Bacteria and fungi in soil

46   Which two carnivores compete for the same food sources?

47   Based on this diagram, which organism is best described as an omnivore?

48   Why are the grasses and shrubs considered "producers?"

49   If an insecticide was used to kill the grasshoppers what would happen to the population of small birds?

Question 50 refers to the table below

LENGTH OF DAYLIGHT FOR THREE CITIES IN THE U.S.

City/ Latitude	May 1 Length of day (Hr: Min)	June 1 Length of day (Hr: Min)	July 1 Length of day (Hr: Min)
Atlanta 25° N	13:35	14:21	14:23
New York 40° N	13:53	14:49	14:58
Boston 45° N	14:15	15:02	15:14

50 As you travel north from Atlanta to Boston in the month of June, what change in the length of daylight will you observe?

_____

Base your answer to question 51 on the following diagram of Cell A and Cell B.

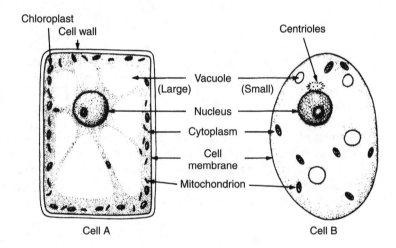

51 Which two structures are typically found in plant cells, but not in animal cells?

_____

Base your answer to question 52 on the following diagram that illustrates a type of reproduction.

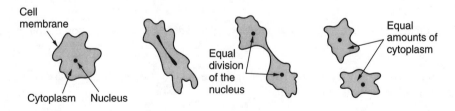

52 What type of reproduction is illustrated in the diagram?

_____

53 Which type of electromagnetic wave has a wavelength of about 10^2 meters and a frequency of 10^6 hertz?

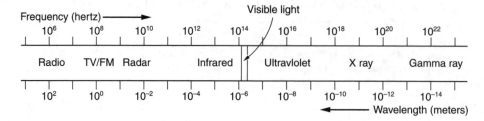

54 List two energy transformations that occur when someone rings a bell

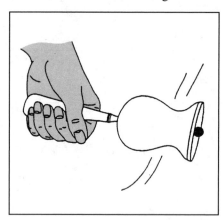

Base your answers to questions 55–59 on the information below.

Midge performed an experiment to determine what concentration of salt water is best for growing a salt-water plant. She prepared five solutions of salt water in different concentrations as indicated in the data table below. She also used pure water in one of her experimental groups. All the plants were the same height when she started. The plants were allowed to grow for 30 days under the same conditions of light and temperature.

At the end of the 30 days, Midge measured the height of each plant and recorded it in the data table below.

Concentration of Salt (%)	Height (cm)
0 (pure water)	2
1	9
2	14
3	17
4	17
5	9
6	1

55 Use the numbered grid on the answer paper to plot the data for this experiment. Use an X to mark each point and connect the points with a solid line.

56 Label the x axis with an appropriate title and proper units.

57 Label the y axis with an appropriate title and proper units.

58   Midge did research and found that the average concentration of salt in the ocean is 3.5%. What conclusion could you draw about the growth of this plant at a concentration of 3.5%?

59   Which conclusion can be drawn correctly from the results of this experiment?

60   In two or more sentences, explain the differences between a mixture and a compound. Give two examples of each.

Base your answers to questions 61 and 62 on the Planetary Data Table. This table shows the average orbital speed and average distance from the sun for the nine planets in our solar system.

PLANETARY DATA TABLE

Planet	Average Orbital Speed (kilometers/second)	Average Distance from the Sun (kilometers)
Mercury	47.60	57,900,000
Venus	34.82	108,200,000
Earth	29.62	149,600,000
Mars	23.98	227,900,000
Jupiter	12.99	778,000,000
Saturn	9.58	1,427,000,000
Uranus	6.77	2,871,000,000
Neptune	5.41	4,497,000,000
Pluto	4.72	5,913,000,000

61   Describe the relationship between a planet's average orbital speed and its average distance from the sun.

62   If a new planet were discovered between Neptune and Pluto, what would its approximate average orbital speed be.

Base your answers to questions 63–65 on the diagrams below. Diagram I represents the moon orbiting Earth as viewed from space above the North Pole. The moon is shown at eight different positions in its orbit. Diagram II represents the phases of the moon as seen from Earth when the moon is at position 2 and at position 4.

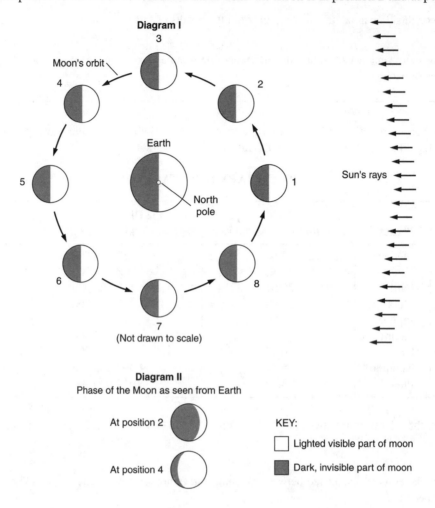

Diagram I

Moon's orbit

Earth

North pole

(Not drawn to scale)

Sun's rays

Diagram II
Phase of the Moon as seen from Earth

At position 2

At position 4

KEY:

☐ Lighted visible part of moon

■ Dark, invisible part of moon

63 Shade the circle provided on your answer paper to illustrate the moon's phase as seen from Earth when the moon is at position 7.

64 State the two positions of the moon at which an eclipse could occur.

65 State the approximate length of time required for one complete revolution of the moon around Earth.

Base your answers to questions 66-70 on the following information.

After learning that zinc metal reacts with hydrochloric acid to produce hydrogen gas, the students in Mr. Jackson's chemistry class designed and performed the following experiment.

The students placed equal masses of zinc into each of four separate test tubes, that were labeled A, B, C, and D. To test tube A, they added 10.0 mL of a 4% solution of hydrochloric acid. Test tube B, received 10.0 mL of an 8% solution of the acid. Test tube C, received an equal amount of a 12% solution, and test tube D, an equal amount of a 20% solution of the acid.

The students measured the amount of time needed to collect 25.0 mL of the hydrogen gas produced from each test tube in the experiment. It took 64 seconds to collect the hydrogen from test tube A. It took 33 seconds from B, 16 seconds from C, and 4.0 seconds to collect the hydrogen from test tube D.

66 Create a data table that clearly and logically summarizes the results of this experiment. Be sure to include a proper heading for each column of your table. Fill in your table with the data from the experiment.

67 What problem was this experiment designed to investigate?

68 Mr. Jackson suggested that the students take the temperatures of each of the acid samples before using them. Why did he make this suggestion?

69 Unfortunately, there was no 16% acid solution available to be tested. Based on the data in your table, predict the time it would take to collect the 25.0 mL of hydrogen using a 16% solution of hydrochloric acid, under the same conditions.

70 What is the relationship between concentration of acid, and the speed of this reaction.

71 Explain why the lungs are considered to be part of both the respiratory system and the excretory system.

72 Describe the role of the skin in human excretion.

Intermediate-Level Science Sample Test 3
Student Answer Sheet for Parts 1 and 2

Student's Name _____

Directions (1–35): Each question is followed by four choices. Decide which choice is the correct answer. Mark your answer in the spaces below by writing the number of the answer you have chosen. (NOTE: A scannable answer sheet will be provided for the actual exam. This format is for the sample test only.)

1 _____	13 _____	25 _____
2 _____	14 _____	26 _____
3 _____	15 _____	27 _____
4 _____	16 _____	28 _____
5 _____	17 _____	29 _____
6 _____	18 _____	30 _____
7 _____	19 _____	31 _____
8 _____	20 _____	32 _____
9 _____	21 _____	33 _____
10 _____	22 _____	34 _____
11 _____	23 _____	35 _____
12 _____	24 _____	

Directions (36–72): For each question, write your answer in the spaces provided below.

36 _____

_____

37 _____

_____

38 _____

_____

39 _____

_____

40 _____

_____

41 _____

_____

_____

42

	y	y
Y	Yy	
Y		

43 _____

44 _____

_____

45 _____

_____

46 _____

_____

47 _____

_____

48 _____

49 _____

_____

50 _____

_____

51 _____

52 _____

53 _____

54 _____

_____

55–57

58 _____

59 _____

60 _____

61 _____

62 _____

63

64 _____

65 _____

66

67 _____

_____

68 _____

_____

69 _____

_____

70 _____

_____

71 _____

_____

_____

_____

72 _____

_____

_____

_____

Glossary

A

acceleration The rate of change in velocity.

acid rain Rain that has been made more acidic than normal by pollutants in the atmosphere.

adaption A characteristic that helps an organism survive in its habitat.

air mass A large body of air that has roughly uniform temperature and humidity throughout.

air pressure The force with which air presses down on Earth's surface.

air temperature A measurement of the amount of heat energy in the atmosphere.

altitude The height above sea level of a place.

amplitude The height of the crest or the depth of the trough of a wave measured from the undisturbed surface.

arteries Blood vessels that carry blood away from the heart.

asexual reproduction Reproduction that involves only one parent, producing offspring that are genetically identical to the parent.

atmosphere The gases that surround Earth.

atom The smallest particle of an element that has properties of that element.

axis of rotation An imaginary line through an object, around which the object spins.

B

bacteria of decay Microorganisms that break down dead organisms and return nutrients to the environment.

bedrock The unbroken, solid rock portion of Earth's crust.

behavioral adaptations Special behaviors that help organisms adapt to changes in their environment.

benefit An advantage to a person or society of a technological device or process.

biodiversity Describes the great variety of life-forms on Earth or within a habitat.

blood A liquid tissue that contains red and white blood cells, and platelets, and also carries dissolved gases, nutrients, hormones, and wastes.

blood vessels Tubes through which the blood flows.

boiling The rapid change in phase from liquid to gas, during which bubbles of gas form within the liquid.

boiling point The temperature at which a substance changes rapidly from a liquid to a gas.

bone The hard parts of the skeleton.

brain The organ, located within the skull, that controls thinking and body activities.

bronchi The two tubes that branch off from the lower end of the trachea, connecting it to the lungs.

393

buoyancy The tendency of an object to float.

buoyant Able to float.

burden A disadvantage to a person or society of a technological device or process.

C

Calorie 1000 calories; used to measure the heat energy in foods.

calorie A unit used to measure and compare the amount of heat energy contained in substances such as fuels.

cancer A disease that is the result of abnormal cell division.

capillaries Tiny blood vessels, connecting arteries to veins, through which materials are exchanged between the blood and the body's cells.

carnivore A meat-eating animal.

cartilage A flexible tissue that acts as a cushion between bones and provides flexibility at the ends of bones.

cell The basic unit of all living things.

cell division The process by which cells reproduce, wherein a parent cell splits into two new daughter cells.

cell membrane The outer covering, or "skin," of a cell, which controls the flow of materials into and out of the cell.

cellular respiration A life process that occurs in all cells, in which nutrients from digested food are combined with oxygen to release energy and produce the wastes carbon dioxide and water.

chemical bond The link that joins one atom to another in a molecule.

chemical change A change that results in the formation of one or more new substances; a chemical reaction.

chemical property A characteristic that a substance displays when it undergoes a change to a new substance or substances.

chlorophyll A green substance found in the leaves of plants, which is necessary for photosynthesis to take place.

chloroplast The organelle in plant cells that contains chlorophyll.

circuit breaker A device that prevents overloading of an electric circuit by interrupting the flow of electricity when it reaches a dangerous level.

classification system A system that groups things together by properties, which you choose.

climate The general character of the weather in an area over many years.

cloud A mass of tiny water droplets or ice crystals suspended high in the atmosphere.

coal A black rock formed from the remains of ancient swamp plants; it is a fossil fuel.

coarse-adjustment knob The part of a microscope used to focus under low power.

cold front The boundary formed when a cool air mass pushes into and under a warm air mass.

community All the different populations that live within a habitat.

compound A substance that is formed when two or more different elements combine chemically.

compound microscope A microscope that uses two lenses.

condensation The changing of water vapor into droplets of liquid water; more generally, the change in phase from gas to liquid.

conductor A material through which electricity can flow.

conservation The saving of natural resources through wise use.

constellation A pattern formed by stars in the night sky, which reminded people of animals or characters in ancient myths.

consumer An organism that obtains nutrients by eating other organisms.

contour line A line on a topographic map that connects points of equal elevation.

core Earth's center, which is made up of an outer and inner zone.

corrosion The chemical wearing away of a metal.

crust The outermost rock layer of Earth, which contains all of Earth's surface features.

cytoplasm The watery substance that fills the cell, where most life processes occur.

D

decomposer An organism that breaks down the remains and wastes of other living things.

density The quantity that compares the mass of an object to its volume $density = \frac{mass}{volume}$.

digestion The breaking down of the nutrients into a useable form.

Doppler effect The apparent change in the frequency of a wave (light or sound) that occurs when the source and/or the observer are in motion relative to each other.

dormancy A state in which an organism is inactive while it awaits more favorable conditions in its environment.

E

earthquake A shaking or vibrating of Earth's crust, usually caused by the sudden movement of rocks sliding along a fault.

ecological succession The natural process by which one community of living things is replaced by another community in an orderly, predictable sequence, until a stable climax community appears.

ecosystem The living members of a community, plus the nonliving elements of their environment.

effort The force applied to a simple machine to overcome the resistance.

electric circuit A complete path for the flow of electricity.

electric energy A form of energy produced by the flow of electrons from one point to another point.

electromagnetic induction The process of using a magnet to induce a flow of electrons in a coil of copper wire.

electromagnetic waves Energy waves that travel at the speed of light and can move through a vacuum; they include radio waves, microwaves, infrared waves, visible light, ultraviolet waves, X rays, and gamma rays.

element One of the basic substances that form the building blocks of matter.

elimination The removal of undigested materials from the body (egestion).

endangered species A species that is in danger of extinction.

energy The ability to do work.

environment The surroundings in which an organism lives, including both living and nonliving things.

erosion The process whereby rock material at Earth's surface is removed and carried away.

evaporation The changing of liquid water into water vapor (gaseous water); more generally, the change in phase from liquid to gas.

evolution The process by which a species gradually becomes a different species.

excretion The process of removing the waste products of cellular respiration from the organism.

extinct Species that no longer exist.

F

family A subgroup of an order.

faulting The process in which internal forces cause Earth's crust to break and slide along a fracture called a fault.

fertilization The joining of an egg cell and a sperm cell, during sexual reproduction, to begin the development of a new individual.

field of view The area you see through the microscope.

fine-adjustment knob The part of a microscope used to focus under high power.

first law of motion An object at rest will remain at rest and an object in motion will remain in motion unless an outside force acts on the object.

folding The process whereby rock layers in Earth's crust are squeezed into wavelike patterns called folds.

food chain A sequence of organisms through which nutrients are passed along in an ecosytem.

food web A number of interconnected food chains.

force A push or a pull.

fossil The remains of traces of an ancient organism.

fossil fuel A fuel that was formed from the remains of ancient plants or animals; examples include oil, coal, and natural gas.

freezing (1) The change in phase from liquid to solid. (2) The storing of food at temperatures below 0°C (32°F), to slow the growth of microorganisms that can spoil food.

freezing point The temperature at which a substance changes from a liquid to a solid.

frequency The number of waves that pass by a fixed point in a given amount of time.

front The boundary between two different air masses.

full moon The phase of the moon that occurs when Earth is between the sun and the moon, so that all of the moon's lighted side can be seen from Earth.

fuse A device, used in an electric circuit, containing a thin metal strip that melts to interrupt the flow of electricity when the circuit becomes overheated.

G

gene A piece of genetic information that influences a trait.

genetic engineering The process of changing the genetic material of organisms.

gland An organ that makes and secretes (releases) chemicals called hormones.

greenhouse effect The trapping of heat in Earth's atmosphere by carbon dioxide, can lead to global warming.

growth The increase in size of an organism.

H

habitat The particular environment in which an organism lives.

heart An organ, made mostly of muscle, that contracts (beats) regularly to pump blood throughout the blood.

heat energy The energy of motion of the vibrating particles that make up matter.

herbivore A plant-eating animal.

hibernate To enter a sleeplike state of reduced body activity; how some animals survive the winter.

hibernation A sleeplike state of reduced body activity that some animals enter to survive the winter.

high-pressure system A large area where air is sinking, causing high surface air pressure; also called a high.

hormone A chemical "messenger" secreted by a gland into the bloodstream, which carries the hormone to an organ that responds in some way.

humidity The amount of moisture (water vapor) present in the atmosphere.

hurricane A huge, rotating storm that forms over the ocean in the tropics, with strong winds and heavy rains.

hydroelectric energy Electricity produced by using the energy of flowing water to turn the turbines of a generator.

hydrosphere The liquid part of Earth.

I

igneous rock A rock formed by the cooling and hardening of hot, liquid rock material.

inclined plane A simple machine that consists of a flat surface with one end higher than the other, such as a loading ramp.

inertia The tendency of an object at rest to remain at rest or an object in motion to remain in motion.

infectious disease A disease caused by microorganisms that can be transmitted from one individual to another.

ingestion The process of taking in food.

inner core Earth's center, it has a radius of 1200 kilometers, and it is thought to be solid because P-waves travel faster through it.

insoluble Not able to dissolve in a given solvent.

insulation Material used to reduce or slow the flow of heat from one area to another.

insulator A material through which electricity cannot flow.

involuntary muscles Muscles that we do not consciously control.

J

joint A place where one bone is connected to another one.

K

kidneys A pair of organs that filter wastes from the blood and help control the water and mineral balance of the body.

kinetic energy Energy that an object has because of its motion.

kingdom The largest group in Linnaeus' classification system for living things.

L

latitude Distance from the equator, measured in degrees.

Law of Conservation Energy Energy can neither be created nor destroyed.

Law of Conservation of Matter Matter can neither be created nor destroyed in a chemical reaction.

lens A piece of transparent glass or plastic with curved surfaces that bend light rays to form an image.

lever A simple machine consisting of a bar or rod that can turn around a point called the fulcrum.

life cycle The changes that an organism undergoes as it develops and produces offspring.

light A visible form of radiant energy that moves in waves.

light-year The distance that light travels in one year, about 9.46 trillion kilometers.

lithosphere The solid part of Earth.

liver An organ that produces urea from excess amino acids, removes harmful substances from the blood, and secretes bile, a digestive juice.

locomotion The movement of the body from place to place.

longitude The distance measured in degrees east and west of the prime meridian.

longitudinal wave (L-wave) An earthquake wave that travels along Earth's surface.

low-pressure system A large area where air is rising, causing low surface air pressure; also called a low.

lungs A pair of organs, located in the chest, that contain millions of tiny air sacs, in which the exchange of respiratory gases between the blood and the environment takes place.

lymph A fluid that bathes all body cells and acts as a go-between in the exchange of materials between the blood and the cells.

lymph vessels Tubes in which waste-laden lymph is collected and returned to the bloodstream.

M

machine A device that transfers mechanical energy from one object to another object.

magnification The number of times the image of a specimen is enlarged.

mammary glands The female breasts, which produce milk to nourish newborn offspring.

mantle The layer below the crust.

mass The amount of matter in an object.

matter Anything that has mass and takes up space.

melting The change in phase from solid to liquid.

melting point The temperature at which a substance changes from a solid to a liquid.

metabolism The sum of all the chemical reactions that take place in the body.

metal A shiny solid that conducts electricity. Metals are found to the left on the Periodic Table of the Elements.

metamorphic rock A rock produced when existing igneous or sedimentary rock undergoes a change in form caused by great heat, pressure, or both.

metamorphosis The process of a complete change in body form during development from juvenile to adult stages.

meteor A rock fragment traveling through space that enters Earth's atmosphere and burns up, producing a bright streak of light.

microorganism An organism that is very small, usually too small to be seen with the unaided eye.

microscope A tool used by scientists to magnify tiny objects such as cells.

migrate To move from one environment to another, where conditions are more favorable; how some animals survive the change in seasons.

migration Moving from one environment to another.

mineral A naturally occurring, solid inorganic (nonliving) substance with characteristic physical and chemical properties.

molecule The smallest particle of a compound.

motion A change in the position of an object relative to another object, which is assumed to be at rest.

mountain A feature on Earth's surface that rises relatively high above the surrounding landscape.

muscles Masses of tissue that contract to move bones or organs.

mutation A change in the genetic material.

N

natural gas A gaseous fossil fuel found trapped deep underground, often with oil deposits.

natural selection The process that favors those organisms that are best able to survive and reproduce

nerves Thin strands of tissue, composed of neurons, that carry impulses throughout the body,

neurons Cells that make up the nervous system, which receive and transmit information in the form of impulses.

new moon The phase of the moon that occurs when the moon is between Earth and the sun, so that the moon cannot be seen from Earth.

noble gases A group of gaseous elements that seldom react with other elements, which are placed in the extreme right column of the Periodic Table of the Elements.

noninfectious disease A disease that cannot be transmitted from one individual to another.

nonmetal Solids and gases found at the right of the Periodic Table of the Elements, which are poor conductors of electricity

nonrenewable resource A resource that is not replenished by nature within the time span of human history.

nuclear energy The energy stored within the nucleus of an atom, used by nuclear power plants to produce electricity.

nuclear waste The poisonous, radioactive remains of the materials used to fuel nuclear power plants.

nucleus (1) The structure within the cell that controls cell activities and contains genetic material. (2) The center of an atom.

nutrients Food substances that supply an organism with energy and with materials for growth and repair.

nutrition The process that includes ingestion, digestion and elimination.

O

oil A thick, black, liquid fossil fuel, found trapped underground; also called petroleum.

omnivore A consumer that can eat both plants and other animals.

orbit The path of an object in space that is revolving around another object.

organ A group of tissues that act together to perform a function.

organism A living thing.

organ system A group of organs that act together to carry out a life process.

outer core Surrounding the inner core, it is about 2300 kilometers thick. It is thought to be liquid because S-waves cannot travel through it.

ovaries The female reproductive organs that produce egg cells.

oviducts Tubes that connect the ovaries to the uterus.

P

parallel circuit An electric circuit that has two or more paths for the electricity to flow through.

phases (1) The changing apparent shape of the moon, as seen from Earth. (2) The three forms, or states, of matter—solid, liquid, and gas.

photosynthesis The process by which green plants produce food, using sunlight, carbon dioxide, and water; oxygen is given off as a by-product.

physical adaptation A physical characteristic that enables an organism to survive under a given set of conditions.

physical change A change in the appearance of a substance that does not alter the chemical makeup of the substance.

physical property A characteristic of a substance that can be determined without changing the identity of the substance.

plain A broad, flat landscape region at a low elevation, usually made of layered sedimentary rocks.

plateau A large area of Earth's surface made up of horizontally layered rocks, found at a relatively high elevation.

plate tectonics The theory that Earth's crust is broken up into a number of large pieces, or plates, that move and interact, producing many of Earth's surface features.

pollutants Harmful substances that contaminate the environment, often produced by human activities.

population All the members of a particular species that live within a habitat.

potential energy Stored energy that an object has because of its position or chemical makeup.

precipitation Water, in the form of rain, snow, sleet, or hail, falling from clouds in the sky.

prevailing winds The winds that commonly blow in the same direction at a given latitude.

primary wave (P-wave) An earthquake wave that can travel through liquids and solids.

producer An organism that makes its own food. Most producers are green plants.

profile The side view of a landform projected from a straight line on a contour map.

pulley A modified form of a lever that can be used to change the direction of force or decrease the force used to move a heavy object.

Punnett square A diagram used to predict the probability of an organism inheriting a given trait.

R

regulation The process that helps an organism maintain a constant internal environment.

renewable resource A resource that is replenished by nature within a relatively short time span.

reproduction The life process by which organisms produce new individuals, or offspring.

resistance The force a machine has to overcome.

respiration (1) The process of taking in oxygen from the environment and releasing carbon dioxide and water vapor. (2) See also cellular respiration.

response The reaction of a living thing to a change in its environment.

revolution The movement of an object in space around another object, such as the revolution of the moon around Earth.

rock A natural, stony material composed of one or more minerals.

rotation The spinning of an object around its axis.

S

science The study of the natural world.

screw An inclined plane that is wrapped around a wedge or cylinder.

seafloor spreading New rock material upwelling along the ridge and moving east and west away from the ridge in a conveyor beltlike fashion, pushes the seafloor out in opposite directions.

secondary wave (S-wave) An earthquake wave that can only travel through solids.

second law of motion The relationship among force, mass, and acceleration: $F = m \times a$.

sedimentary rock A rock formed from layers of particles, called sediments, that are cemented together under pressure.

selective breeding A process in which individuals with the most desirable traits are crossed or allowed to mate with the hope that their offspring will show the desired traits.

sense organs Organs that receive information from the environment. The sense organs include the eyes, ears, nose, tongue, and skin.

series circuit An electric circuit that has a single path of electricity through the devices attached to it.

sexual reproduction Reproduction that involves two parents, producing offspring that are not identical with either parent.

skin The organ that covers and protects the body, and excretes wastes by perspiring.

smog A haze in the atmosphere produced by the reaction of sunlight with pollutants from cars and factories.

soil A mixture of small rock fragments and decayed organic material that covers much of Earth's land surface.

solar energy Energy from the sun.

solar system The sun and all the objects that revolve around it, including the planets and their moons, asteroids, comets, and meteors.

solubility The maximum amount of solute that can dissolve in a given amount of solvent at a given temperature.

soluble Able to dissolve in a given solvent.

solute The substance that dissolves in the solvent.

solution A mixture in which the components remain evenly distributed.

solvent The substance that dissolves the solute.

sound A form of energy produced by a vibrating object.

sound waves Alternating layers of compressed and expanded air particles that spread out in all directions from a vibrating object.

species A group of organisms of the same kind that can produce fertile offspring. The subgroups of a genus.

specimen The object to be viewed.

speed The distance traveled per unit of time, for example, meters per second.

sperm ducts Tubes through which sperm cells pass upon leaving the testes.

spinal cord The thick cord of nerve tissue that extends from the brain down through the spinal column.

stimulus A change in the environment that causes an organism to react in some way.

storm A natural disturbance in the atmosphere that involves low air pressure, clouds, precipitation, and strong winds.

succession See ecological succession.

symbiosis A relationship in which two or more different organisms live in close association with one another; that is, when one organism lives on or inside another one.

system A group of related elements or parts that work together for a common purpose.

T

technology The application of scientific knowledge and other resources to develop new products and processes.

testes The male reproductive organs that produce sperm cells.

thermal pollution An increase in the temperature of a body of water, caused by human activities, that may be harmful to living things in that environment.

third law of motion For every action there is an equal and opposite reaction.

thunderstorm A brief, intense rainstorm that affects a small area and is accompanied by thunder and lightning.

tides The rise and fall in the level of the ocean's waters that take place twice each day.

tissue A group of similar cells that act together to perform a function.

tornado A violent whirling wind, sometimes visible as a funnel-shaped cloud.

trachea The tube that connects the nose and mouth to the bronchi, which lead to the lungs; also called the windpipe.

transport The process of moving materials throughout an organism.

U

uranium A radioactive element found in certain rocks and used as a fuel for nuclear power plants.

uterus The organ of the female reproductive systems within which the offspring develop; also called the womb.

V

vagina The birth canal.

veins Blood vessels that return blood to the heart.

velocity The speed of an object in a certain direction.

volcano (1) An opening in Earth's surface through which hot, liquid rock flows from deep underground. (2) A mountain formed by a series of volcanic eruptions.

volume The amount of space an object occupies.

voluntary muscles Muscles that we consciously control.

W

warm front The boundary formed when a warm air mass slides up and over a cool air mass.

water cycle The process in which water moves back and forth between Earth's surface and the atmosphere by means of evaporation, condensation, and precipitation.

watt A unit that measures the rate at which electrical energy is used.

wavelength The distance from one point on a wave to the corresponding point in the next wave.

weather The changing condition of the atmosphere, with respect to heat, cold, sunshine, rain, snow, clouds, and wind.

weather forecasting An attempt to make accurate predictions of future weather.

weathering The breaking down of rocks into smaller pieces.

wedge A double-sided inclined plane.

wheel and axle A modified lever that consists of a large wheel and a smaller wheel, or axle, at its center.

wind The movement of air over Earth's surface.

wind direction The direction from which the wind is blowing.

winter storms Blizzards and ice storms.

work The moving of an object over a distance by a force.

Index

Photo Credits: